Pico RP2040 Microcontroller and Python

DIY projects

Christopher Dock

Books by Christopher Dock

Getting started with Arduino and Raspberry pi

Smart Picture Frame Raspberry Pi DIY projects

Making your first million publishing e-books

Mechanical keyboards from switches to keycaps

It wasn't me ... Tales of mayhem at work

You wouldn't believe what happened at work today ... tales of the world

Lets go with that! Tales of the world

Unbelievable workplace stories from the planet earth

Modern Travel ... the Nightmares Keep on Coming

ISBN: 978-1-952930-18-8

North American Offices
Christopher Dock
3533 Rum River Drive
Anoka, MN 55303
USA

European Offices
Christopher Dock
Ludwig-Richter-Straße 15
60433 Frankfurt
Germany

DEDICATION

For my family whose patience and encouragement made this book possible, and Covid 19 which gave me extra time at home to work on projects.

CONTENTS

ACKNOWLEDGMENTS

Developing most anything tends to be a group effort and writing a book is no different. I was lucky enough to receive support from my family and friends to whom I am eternally grateful.

1 INTRODUCTION

Small powerful single board computers that run their own Linux operating system are amazing. The Raspberry Pi 4 is perhaps more powerful than network servers from a decade ago all the while costing less than 100 dollars. They support a full sized multi-tasking Linux operating system with the incredible software library. There is virtually no limit to how these small computers can be used. They can run Samba for creating a device that can share storage, Apache for running a web server, CUPS if you wish to share a printer or even create a proxy server for your home network.

The Raspberry Pi v4 is the most recent and capable incarnation of this line of small computers. It can be purchased with 2mb, 4mb or even 8mb of ram but still be small enough that it can fit into your pocket or purse. There is almost no downside to this computer outside of cost. Initially you could easily purchase this computer for 35 USD from quite a few vendors.

Perhaps due to its success and the pandemic situation the price[1] has only gone up[2] and but to add insult to injury lately[3] it can usually only be purchased as a package with other peripherals.

A Raspberry Pi is a great addition to many projects if the project requires intensive calculations or internet connectivity. However, it can be overkill in both cost and capability for many tiny projects. Not only that it has the same weakness that all computers have, it takes a while to start up and needs to be carefully shutdown – but there is a solution.

Raspberry Pi Pico

The latest release in the Raspberry pi family is the Raspberry Pi Pico. The Pico is a small board powered with a system on chip(SOC). A SOC contains many of the dedicated parts that you normally find separately on a computer motherboard but all conveniently in a single chip. The SOC typically has a CPU, GPU, as well as special processing units or communication functionality. SOC devices are used in specialty situations where a full blown computer is not feasible.

The Pico is not a revolutionary device but it is the most powerful while simultaneously the easy to use micro-controller device at the

1 To be fair, the capabilities have also increased since the initial release of the Raspberry Pi.

2 It is difficult to purchase a Raspberry Pi v4 due to stock issues and the price point has moved to approximately 150 USD.

3 June 2023

moment.

This goal of this book is to provide enough information for anyone interested in enhancing their own small projects with an electronic component.

Navigating through the book

The content in this book was created in order for someone with little or no knowledge with a Raspberry Pi Pico. The idea was you can start at page 1 and work your way through. However, depending on your interest and as well as previous computing experience it is possible to skip over some sections you are already familiar with or less interested in.

Below is a summary of what each chapter will cover. They can essentially be read in any order but the later chapters do assume some knowledge from the earlier chapters.

Chapter 1 - Introduction
Chapter 2 - CPU's and microcontrollers
Chapter 3 - Setting up Pico environment
Chapter 4 - Python and Micropython
Chapter 5 - Thonny Development Env
Chapter 6 - First MicroPython program
Chapter 7 - Smiling hardware
Chapter 8 - Developing with C/C++

Assumptions

This book covers a lot of different information about the Pico. The assumption is that the reader has a low to no experience with electronics or small micro electronics but perhaps some experience with computers and software development.

This book is starting from a beginners level and working through form setup of your programming environment through to the actual programming of your Pico to do "things" or to control external components or devices. Enough information is provided that a novice can get started with any of the topics in this book. This will allow you to become familiar with Pico or use it in the scope of a larger DIY project. With some extra effort it is possible to go beyond this micro processor platform and create your own device or IOT device

Despite this book telling a narrative from the "unboxing" through to completed device it is not a fantasy novel you would read while laying at the beach. This book can be used as a guide and to a certain degree as a reference but despite all the information covered it cannot answer every question sometimes more information is necessary.

Additional information might be found in other books but more and more often it is found on the internet, provided by parts vendors or even in the form of a video on Youtube or from other hobbyist sites.

Conventions

The following conventions have been used in this book.

`constant width`	Code examples, fragments, as well as other text related to the code or the building thereof.
Bold	Is used in examples to demonstrate command that must be typed exactly
italic	Is used for displaying the output from commands or programs.
$	Is the prompt of the Bash shell where commands are typed.
#	Is the prompt from the bash shell where the root user is logged in.
c:\	Is the windows command prompt from where commands are typed.

2 CPU'S AND MICROCONTROLLERS

Just about everyone knows what a computer is and to some degree how they operate. You double click on an icon or select a program from the menu and within seconds your screen displays its user interface for you to react to. Your processor controls the flow of data whether they be network requests, calculations or directives for your graphics card.

The CPU is a complex transistor but it gains its strength by working with other single purpose devices such as motherboard, graphic card and storage to create an overall general purpose device. The CPU itself does not need to be shutdown in an orderly manner but the rest of your operating system does. This is because some data might have not have been written to the disk. It is the orderly shutdown that ensures this is done to prevent information from being lost.

This is not too much of a hardship

considering we don't expect computers to turn on instantly and most of the tasks we perform usually take more than a few seconds to complete. It is important that our tasks are saved and so an organized shutdown of our computer is necessary step. Does it really have to be this way?

The short answer is no. The high tech world that we live in has plenty of examples where we can turn on and off electronic items with no notice and with no adverse side effects. Some everyday examples are in our household.

dvd player
microwave
stoves
dishwasher
refrigerators
electric toothbrush
washing machine
vacuum cleaner
printers

This is only possible because these devices are not powered by general purpose computer but rather by a micro-controller.

Micro-controller

A micro-controller is a special purpose chip which is programmed for primarily for a single task while a computer CPU has no fixed task and has the power to process large amounts of data.

	CPU Intel I5 12th Gen	Micro-controller Atmega 328
Speed	3.3 - 4.4 ghz	16 mhz
Local cache / memory	12MB	32k
Interfacing	PCI Bus	I2C, SPI, UART, USB
Cores	8	1
Efficiency Cores	4	0
Power	65W	0.5W
Cost	198 Eur	3.45 Eur

Listing 1 computer CPU vs micro-controller

The differences of the two are different chips could not be greater, one is a powerhouse capable of processing terabytes of data, while the other sips power and can control many devices over one of the specialized communication protocols such as I2C, SPI or UART.

I2C

In the 1980's Philips Semiconductors developed the I2C bus to allow their I/O components in their audio video equipment to communicate with each other. Previous to this components would have direct connections to each other. This is acceptable for a small number of components but it balloons out of control as the number of components that needs direct connections increases. The number of direct paths for three components is six while the number of paths for six components is thirty.

$$connections = \frac{components \times (components - 1)}{2}$$

The I2C bus solution was a huge step forward. All components would need to connect to a common bus, have a common clock and would be able to communicate with each other. The cost for adding one more component would be simply connecting it to the existing bus.

The bus was composed of two wires. One of the wires is used for the data(SDA) while the other is used for the clock signal(SCL). Each device on the bus would have its own unique address. The address is a 7 bit number which limits the number of I2C devices that can be used to 128. Each I2C device[4] will listen to the data line to see if any commands are being passed to for the address it is assigned to.

Any circuit that uses the I2C protocol will have one master and one or more slaves that the master will communicate with. Only the master can initiate a transfer over the bus. Once some command has been initiated the slave can respond. The original patent on this protocol expired in 2004; because the patent has expired there are other implementations that are sometimes called "2 wire" but is actually the same. This protocol is actually directly supported by quite a few integrated circuits and chips.

Advantages

- Uses two wires
- ACK/NACK confirmation for flow control
- Hardware less complicated than UART
- Well known protocol

4 Slave device

Disadvantages

Slower transfer rate than SPI

More complicated than SPI to implement

Transfers 8 bits at a time

SPI

It wasn't only Philips in the 1980's who were working on short distance communications between components. Motorola created their own Serial Peripheral Interface (SPI) protocol for transferring data and communicating between components.

In some ways, the SPI protocol was similar to the I2C protocol. It also had a master/slave relationship where the master controls which slave it communicates with. Unlike the I2C protocol the SPI has four lines that connect the master with the slave. Instead of an address of the slave component it has a direct control using the slave using the chip select(CS) line. It controls if the chip is active by bringing this logic line to 0. Communication with the is done over the Master Out Slave In (MOSI) and the Master In Slave Out (MISO) lines. The master will communicate to the selected slave by sending data over the MOSI line. The responses of the slave will be transmitted back to the master over the MOSI line. The communication is kept in sync with the fourth line being a clock line.

SPI is sometimes referred to as a "4 wire" protocol. Although more lines are used in this protocol to communicate between components it tends to be simpler. Neither the master nor the slaves need to listen to the data line to see if commands or responses are intended for them. The master knows which component it wishes to control and activates it accordingly, while the this same activation enables the chip.

Although it tends to be very convenient to transmit 8 bits between the master and slave, this 8 bit communication is not limited to 8 and could just as easily be 16. These two features alone reduce the amount of overhead when transferring data but the fact that there is two lines (MOSI, MISO) it is possible for the the master and slave to communicate with each other simultaneously. These features can allow for higher transmission of data between the components.

Advantages

- Simpler software implementation
- Faster protocol
- Flexible, not limited to 8 bit transfers
- Slaves not limited by 7 bit address
- One unique line per client, the rest are shared
- Uses fewer lines than parallel data transfer

Disadvantages

- Uses four wires
- requires more pins on micro-controller (slave count +3)
- No error checking protocol

UART

The UART is typically referred to as serial communication. It was first developed by Western Digital in 1971. Unlike the SPI and I2C protocols the UART was designed for connecting two devices together. The UART uses four wires for the transmission of the data.

VCC
Ground
TX (transmit)
RX (receive)

The data was transmitted one bit at a time sequentially, but this often includes other protocol bits to ensure the syncing between the source and destination devices. Each character is sent as a series of bits - a start bit, data bits, parity bit and a stop bit.

Single board Raspberry Pi Pico
The Raspberry Pi Pico is an amazing little device. The specs are amazing especially when compared against the original Arduino[5] platform.

Size:	21 × 51 mm
Processor:	Dual-core ARM Cortex-M0+
Clock:	Up to 133 MHz
Memory:	264 kB SRAM, 2 MB QSPI Flash

Ports:

26x multi-function GPIO pins
3x analog inputs
2x UARTx
2x SPI controllers

[5] It is also quite difficult to decide exactly which Arduino the Pico should be compared against. There are at least 19 Arduino devices along with untold number of compatible devices created third party vendors.

2x I2C controllers
16x PWM channels

Input power: 1.8-5.5 V DC

It is not a completely fair comparison as the Pico is sporting a state of the art RP2040 micro-controller[6] at 133 MHz instead of a simple micro-controller from a few years back. This system-on-a-chip, SOC, combines all of the communications that a typical micro-controller would have merged together with a low end CPU.

The RP2040 chip name actually encodes information about the capabilities of the chip. Decoding this name is somewhat obvious when for the number of cores but the codes for ram is not quite as obvious.

```
RP  2  0  4  0

^   ^  ^  ^  ^
|   |  |  |  |
|   |  |  |  floor(log2(nonvolatile / 16k))
|   |  |  |
|   |  |  floor(log2(ram/16k))
|   |  |
|   |  Type of core(ie M0+)
|   |
|   Number of cores
|
Raspberry Pi
```

The Pico board format style outwardly looks similar to a very large "DIP" style chip. That is to say with two rows of pins one on each side of the board.

---

6 dual core Arm Cortex M0+

This is in contrast to the Raspberry Pi computer which took the decision for all GPIO pins in a single block.

Having the pins around the circumference of the board is not completely new design; the Arduino Nano has been using this style for a number of years. One small difference between the Nano and Pico is the "castellations",Figure 1, next to the pin holes. This allows the board to be soldered directly to a carrier board.

Figure 1: Castellations

The Arduino's micro-controller, Atmel is

programmed using C/C++. This a super language for performance and for keeping the size of the executable down. Yet, C/C++ is not notorious for being easy to use whereas other interpreted languages such as Python is certainly more user friendly.

The Pico, unlike many of the other older micro-controller platforms, can be programmed with either MicroPython or C/C++. MicroPython is small open source python interpreter that can be run on a micro-controller or embedded development boards. You can program your tasks in MicroPython, which is somewhat unusual in the micro-controller space, but there is one more feature that really sets the Pico it apart from some of the previous generations micro-controller boards - how to program the device.

The Pico gets mounted as "disk" to your computer. You simply copy the program to the "disk" and it gets installed. This is considerably more convenient than the older method of programming a micro-controller which required an intermediate piece of hardware, Figure 2: Micro-controller programmer. This device is between the micro-controller board and the host computer. The chip is placed into the zero insertion chip holder, the colored data wires are connected to the appropriate chip legs and the programmer itself is then connected to a personal computer.

Figure 2: Micro-controller programmer

Header Pins

The Raspberry Pi Pico has a total of 40 GPIO pins which can be used in projects. What are GPIO pins? Well, GPIO stands for general purpose input/output. These pins are the exciting part of any small form factor board as it allows the developer the opportunity to interact with his or her environment.

It is possible to use these pins to read the state of a switch, pressed or not, as a method to gather user input. These pins can also be used for turning on LED's or switching motors on or off.

The Pico has forty pins but not all of these can be used to control devices. Of these only twenty six can be used while the rest of the pins are used for various other functions such as power and ground. Of the twenty six pins there is a special subset, they are analog pins. These three pins are a bit special as they can be used for reading or writing 16 bit analog values ranging from 0 and 65535. These three analog pins are GP26, GP27, and GP28. This is really useful for interacting with the outside world but there is another creative way for controlling the digital pins other than on and off. The good news is that this problem was both discovered along with its solution "pulse width modulation"(PWM).

Pulse width modulation is a clever way of making digital outputs look a bit more like an analog output. This is done by creating the output appear to be at different amplitudes despite the fact that there are only on and off states.

If this output is intended to be for power, it can be simulated as a lessor amount by turning off the power part of the time. If the power is on 100% of the time then the power would be the maximum value, and likewise if the power is off 100% of the time the value would be zero. If we wanted to simulate 50% power we would have the power on only half of the time.

This can be easily visualized with a small experiment using a traditional light bulb and the wall switch. If we turn the switch on and off every second we see a flickering light but if we switch it on and off very very fast the bulb would appear to be constantly on but a dimmer level.

Finding your pin

The pins on the board are physically numbered from 1 to 40 starting in the upper left increasing by one in counter clockwise direction. It is actually a bit too much to hope that the GPIO pin numbers can be determined by the physical location. Indeed it is more complicated than that.

One reason this will not work is that the actual GPIO pins are numbered starting at 0. Physical pin 1, upper left hand corner, is actual GP0. Not all of the pins are numbered GPIO pins as they have other tasks such as physical pin 3. This pin is a ground pin but so are pins 8, 13, 18, 23, 28, 33 and 38. All of the pins physical pins along with their GPIO names and functions are listed in the Appendix pin functions.

3 SETTING UP PICO ENVIRONMENT

Working on a Raspberry Pi Pico or another micro-controller platform (ie Arduino) is somewhat different when compared to the standard personal computer experience.

Figure 3: Raspberry Pi Pico

Just looking at this new device, Figure 3, should make it fairly obvious. There is no keyboard, hard disk, card reader, and no HDMI port to connect a monitor.

On a personal computer most software development takes place on the computer itself and the resulting program also runs on that same machine. This is not the case for micro-controller solutions. Micro-controllers have limited memory and are quite slow when compared to a personal computer thus a slightly different development setup is needed.

For micro-controller development, all work will take place on a different computer such as a Windows PC, Raspberry Pi 4 or even a Linux workstation. This "development" computer is called the host and the resulting program will be copied over to the micro-controller board also referred to as the target platform. This process, cross platform development[7], is a fairly common topic in software development especially when creating apps for smart phones.

Environment

We will need a set of tools to create our executable This set of tools or build chain will contain various command lines tools as well as an integrated development environment (IDE). Technically the IDE is not a requirement but it will provide additional support to the developer. Software development is a tough enough task and using a IDE can help to lighten the load.

It is difficult to say exactly which tools are required as this is directly tied to the host machine which is entirely the developers

7 This is sometimes referred to as cross compiling, building software on a different platform from where it will be run.

preference.

What you need for Pico Development.

Host (PC, Raspberry Pi)
Keyboard
Mouse
Monitor

It is presumed that most people planning on using a single board computer, such as the Raspberry Pi or Pico, will already have some of the necessary peripherals at hand.

Describing the setup for every possible platform is not reasonable, but the Raspberry Pi[8] and Windows are described.

Windows development env - Python

Windows can be setup using the same tools as would be found on a Linux computer or the Raspberry Pi, however, the process is not done by downloading them via the operating system package manager.

Windows PC's usually do not have any of the required programs installed and they will need to be individually downloaded and installed. The best solution for Micro-Python development on the Pico is the IDE Thonny.

Thonny

All the information about the Thonny IDE can be found at their homepage, https://thonny.org/, including a link for downloading the installation program. The Thonny development

8 Linux users will have a very similar installation to a Raspberry Pi.

environment is under active development so do not be surprised when the download link will redirect you to github[9].

The download link will bring you to the current release page[10]. At the bottom of this page is list of the different downloads which supports Windows 7 through Windows 10 in both 32 bit and 64 bits. The default is ***thonny-4.1.2.exe*** which is a 64 bit binary but is bundled with Python 3.10. This choice would be fine for Windows 8.1 and newer. If the platform is 32 bit then the choice should be thonny-py38-4.1.2.exe which includes the 32-bit Python 3.8. Both of these Thonny executables uses a standard windows installer.

Thonny configuration - Windows

One of the most important settings for our IDE is the interpreter as this allows programs to be run on the Pico directly from the IDE. This is setup from "configure interpreter" from the Run menu.

9 The Thonny project can be found at https://github.com/thonny/thonny.

10 The current release URL link https://github.com/thonny/thonny/releases/tag/v4.1.2 is the most current version, which at the time of this writing was tag 4.1.2

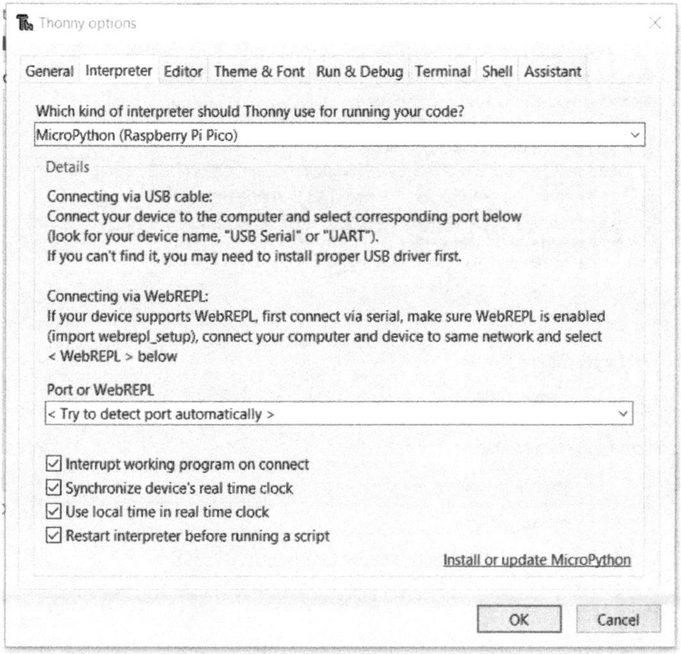

From this dialog, in the lower right hand corner, is a link for installing or updating Micro-Python. Clicking on this will open up another dialog for selecting the Micro-Python.

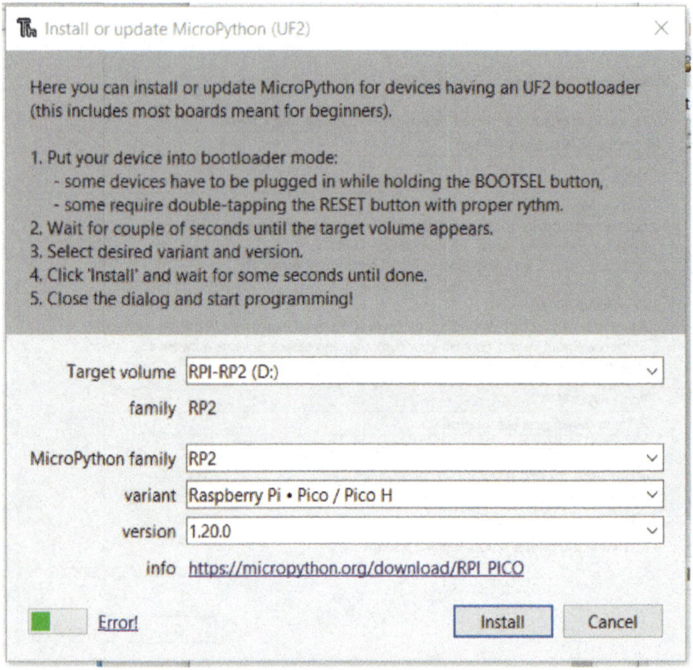

Normally pressing the install button will download and install Micro-Python, but nothing is perfect. In some relatively rare cases an error message will show up instead of Micro-Python being installed.

```
Traceback (most recent call last):
File "C:\Program Files (x86)\Thonny\lib\urllib\
request.py", line 1348, in
do_openh.request(req.get_method(), req.selector,
req.data, headers,
File "C:\Program Files (x86)\Thonny\lib\http\
client.py", line 1283, in
requestself._send_request(method, url, body, headers,
encode_chunked)
File "C:\Program Files (x86)\Thonny\lib\http\
client.py", line 1329, in _send_request
self.endheaders(body, encode_chunked=encode_chunked)
File "C:\Program Files (x86)\Thonny\lib\http\
```

```
client.py", line 1278, in endheaders
self._send_output(message_body,
encode_chunked=encode_chunked)
File "C:\Program Files (x86)\Thonny\lib\http\
client.py", line 1038, in _send_output self.send(msg)
File "C:\Program Files (x86)\Thonny\lib\http\
client.py", line 976, in send self.connect()
File "C:\Program Files (x86)\Thonny\lib\http\
client.py", line 1455, in connect self.sock =
self._context.wrap_socket(self.sock,
File "C:\Program Files (x86)\Thonny\lib\ssl.py", line
513, in wrap_socket return
self.sslsocket_class._create(
File "C:\Program Files (x86)\Thonny\lib\ssl.py", line
1071, in _create self.do_handshake()
File "C:\Program Files (x86)\Thonny\lib\ssl.py", line
1342, in do_handshake self._sslobj.do_handshake()
ssl.SSLCertVerificationError: [SSL:
CERTIFICATE_VERIFY_FAILED] certificate verify failed:
certificate has expired (_ssl.c:1007)
```

Manual install of Micro-Python

If an error occurs it could be a simple message or it could be an entire stack dump. In my case the actual error was "certificate verify failed: certificate has expired".

There have been some reports of this particular error occurring on Windows. Rather than finding and using some cryptic old solution from the internet, we could take an alternative solution. We could simply download and manually install Micro-Python.

There are a few locations that Micro-Python might be downloaded from. The two most obvious are the Micropython homepage or from the Raspberry Pi homepage.

https://micropython.org/download/

The actual source of Micro-Python is from micropython.org and the Raspberry Pi homepage actually will link to the most current version on the micropython.org site. Care must be taken to pick the correct Micro-Python version. There may be multiple different versions could be listed for download as there are three different versions of the Pico available at the moment.

Raspberry Pi Pico
Raspberry Pi Pico W[11]
Raspberry Pi Pico W[12]

The downloaded filename will look somewhat similar to *RPI_PICO-20230426-v1.20.0.uf2* with the date and version being the variable part that may change. Once the file has been downloaded it can be copied to the Pico. The installation process is super simple. When plugging in the Pico to the Windows PC it will be seen as a disk drive. The file should be copied over to the "Pico drive". When a Pico is new and unused it will be automatically mounted as a drive when it is attached to a Windows computer. If the Pico is not automatically mounting as a drive, hold down the white bootsel button when plugging it in and it will then be mounted as a drive. Once the Micro-Python file is copied over, it will be automatically installed and rebooted.

Note: If the Thonny IDE is running during this time, it will need to be restarted before it will recognize the new Micro-Python that is

[11] With WiFi support

[12] With WiFi support and Bluetooth LE support

installed.

Testing Thonny

Once Micro-Python is installed and Thonny is restarted programs can be run directly from the IDE. This can be be tested by running the following program.

```
import time
from machine import Pin
LedPin = Pin(25, Pin.OUT)

for i in range(10):
    LedPin.off()
    time.sleep_ms(100)
    LedPin.on()
    time.sleep_ms(100)
```

This program will cause the Pico onboard LED to blink every 5 times a second for two seconds. Once the test is done, this code can be either discarded or saved. When selecting the Save option from the File menu a dialog will popup asking where to save the file ("this computer" or "Raspberry Pi Pico").

Windows development env - C++

Setting up a windows development for the Pico is a bit trickier. The basic set of tools that will be required is quite small.

ARM GCC compiler
Editor or IDE
Source code control

Picking your software tools is not completely dissimilar to picking a car. There are quite a few that fulfill the requirements in the end the final decision comes down to personal preference. Some people are big fans of open source software due to a fear of vendor lock-in while others like knowing there is a vendor who can help out if necessary.

Rather than attempt to find all of the possible Windows solutions available, I will list a few open source tools or free tools that can be downloaded from the internet. Microsoft also does provide their own enterprise or proprietary solutions.

Compiler

ARM GNU Toolchain	https://developer.arm.com

IDE

Eclipse	https://www.eclipse.org
IntelliJ IDEA	https://www.jetbrains.com/
VSCode	https://code.visualstudio.com

Source control

GIT	https://git-scm.com
Subversion	https://subversion.apache.org/
Mercurial	https://www.mercurial-scm.org/

Build tools

CMake	https://cmake.org/download/

These tools are just a few of the number available in their category. Not all of these tools perform a single task and some may cover multiple different categories. To further complicate things is the fact some of these tools can have their own extensions/plug-ins that modify their standard behavior.

Raspberry Pi development env - Python

The best solution for Micro-Python development on the Pico is the IDE Thonny. The good news is that if Thonny and Micro-Python is not already installed on your Raspberry Pi it can be with just a few commands but there is an even better way. Instead of manually installing and setting up the Raspberry Pi as well as other necessary software, a shortcut that has been provided. This solution was undoubtedly created due to the Raspberry Pi foundation's stated purpose.

> *"further the advancement of education of adults and children, particularly in the field of computers, computer science and related subjects"*[13].

With this goal in mind they have made using both the Raspberry Pi and the Raspberry Pi Pico quite easy to use together. They have created a small shell script that when run on

[13] https://static.raspberrypi.org/files/about/RaspberryPiFoundationStrategy2018%E2%80%932020.pdf

your Raspberry Pi will download, install and setup virtually all the necessary requirements for working with your Pico. Thus you can easily use a Raspberry Pi as your host platform when working on your Raspberry Pi Pico.

This is both in line with the foundations purpose and offers a fairly low barrier to entry for developers.

Raspberry Pi Host Setup

If you already have a Raspberry Pi 3, you don't need to purchase a newer model but the process of getting everything setup might take a bit longer. The actual daily development for the Pico will not be greatly affected by the speed of the host machine.

The first step for preparing your Raspberry Pi is to retrieve the setup script, Figure 4. From a terminal window you will need to do the following command.

```
wget https://raw.githubusercontent.com/rasp\
berrypi/pico-setup/master/pico_setup.sh
```

Figure 4: wget command to get pico_setup.sh script

Note: The for commands that are too long for the page will be broken up with a backslash(\). This should be ignored when typing in the command.

This wget command will retrieve the pico_setup.sh script to the current working directory. It is recommended that this is the home directory (/home/pi) or perhaps in a development directory (/home/pi/development).

Once this script is run it will create a “pico” sub-directory in the current working directory. This can be anywhere but under the Documents, Pictures or Downloads sub-directories it could be considered an odd choice for your development.

Once the pico_setup.sh script has been downloaded it needs to be made executable and then run.

```
$ chmod 755 pico_setup.sh
$ ./pico_setup.sh
```

How long it will take for this script to run depends on your specific situation, the Raspberry Pi 3 runs slightly slower and your internet connection is always a variable factor.

On my Raspberry Pi 4, the entire process, Figure 1, was done in about 10 ½ minutes which included all setup of dependencies, downloading some some examples, compiling a few extra tools as well as preparing the UART for use with the Pico. However, as the last last comment in the process reminds us, we need to reboot the Raspberry Pi in order to finish the UART setup.

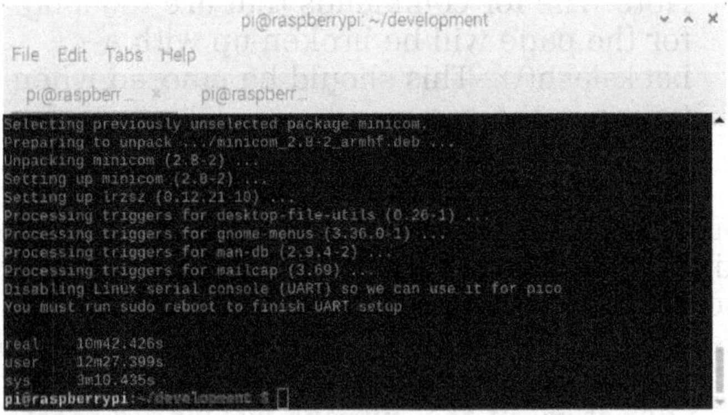

Figure 5: Full run of pico_setup.sh script

Once this script has been run, along with the required reboot, you will be ready to begin development

Additional Raspberry Pi customization

All modern computer systems are designed to work under a given set of conditions. With rare exceptions, you are not required to be working in a climate controlled environment. Depending on the load of the computer it will get warm (or hot) and when it starts to reach its thermal limits it will throttle itself to prevent damage.

It is not necessary to monitor your hardware your Raspberry Pi but I did decide I would watch how it performed while preparing the environment. In order so I could monitor the CPU temperature and memory usage I did a bit of configuration on my Raspberry Pi.

Temperature(optional)

From your desktop graphical environment you can add a temperature app to the panel.

Right click on the top panel
Select "Add / Remove Panel items"
Click "Add"
Scroll down to Temperature Monitor" and select it
Click "Add"
Use the "Up" button to position the monitor above the "Bluetooth"

Adding the temperature app is optional but it is easier to see if it running near the 85C degree temperature. If no measures have been taken for this, cooling fins or fan have been applied to the Raspberry Pi it is isn't a bad idea to keep an eye on the temperature.

Note: The opinion on if you even need either active or passive cooling for a Raspberry Pi is mixed[14]. If the Raspberry Pi does get too hot it will throttle down. There are a lot of affordable neat looking cases that either come with passive fins or an active fan or both.

Graphical Monitor(optional)

It is possible to use the same graphical task monitoring tool that is already installed to your Pi.

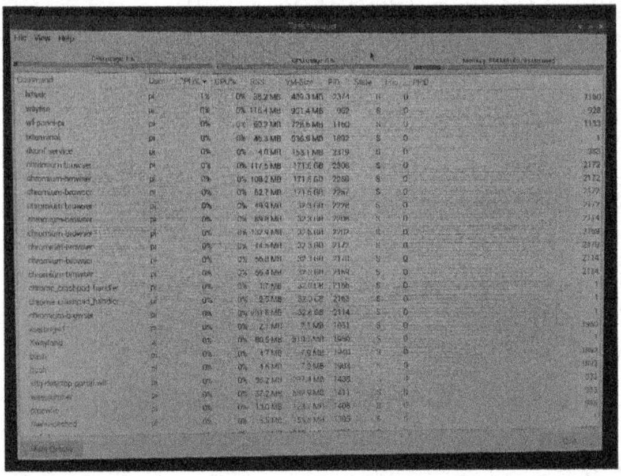

Figure 6: Default process monitor tool

There exists a snap package that can be installed on the Raspberry Pi. Perform the following steps from the command line.

[14] Some opinions on the internet suggest that no active cooling at all is needed for the Raspberry Pi. My experiences have proven this out but if your case is not well ventilated you will be trading cool running for cool looks.

```
sudo apt update
sudo apt install snapd
sudo reboot
sudo snap install core
sudo snap install gnome-system-monitor
```

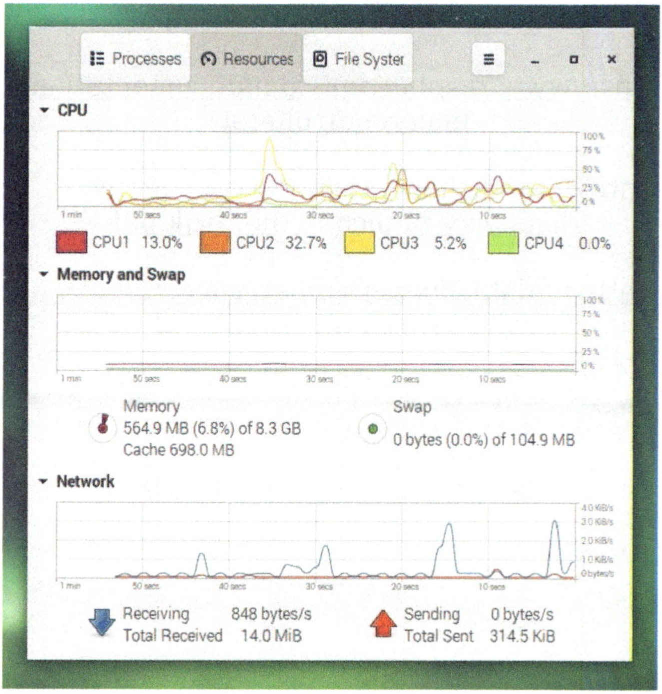

Adding the graphical system monitor takes a few more steps and is less convenient because of the reboot. The executable "gnome-system-monitor" is not automatically added to the Raspberry Pi menus as a separate menu choice but it is used when you select System Monitor from the system tools menu.

Even without these changes optional changes the Raspberry Pi host is ready to be used. The size of the Pico is somewhat misleading, how much technology is inside. For more information refer to the Raspberry Pi Pico documentation.

Raspberry Pi Pico resources page

https://www.raspberrypi.com/documentation/microcontrollers/

https://datasheets.raspberrypi.com/pico/raspberry-pi-pico-python-sdk.pdf

https://datasheets.raspberrypi.com/pico/raspberry-pi-pico-c-sdk.pdf

4 PYTHON AND MICROPYTHON

The python language has actually been around for quite a few decades. It is an interpreted high level language designed to be easily extended.

Not all of what is often considered basic functionality is built into the core language itself but puts this "extra" functionality into a separate module. To use functionality from modules will require that it is explicitly included by each program in the form of an import. Even array support was designed as external module that would need to be imported into your program in order to be used. In addition you can create your own modules separating out specific logic so it can be reused in other programs.

The language supports both object oriented features as well as traditional structured programming. To help with readability and maintainability, like all other languages, your programs should be broken up into functions and methods.

Python strives for elegance, there is typically only one way to perform a given operation, unlike Perl. Python is also rather unique as the blocks of code for the statements are done with indentation instead of with curly braces or other key words.

As an interpreted language it is easy to create scripts for doing small tasks or integrating or automating various operations. Python is a good choice where the speed and performance is not the most important factor. Yet, when performance is important it is possible to use the pypy just in time compiler.

Python Tutorial

Perhaps the title should say a quick tutorial. Python, the core functionality, just like many other languages can probably be picked up in a few hours. Yet in order to solve non-trivial problems will require knowledge of the libraries or modules and that could be a book or books just to fully cover this.

Python programs can be used for big and complex programs but the resources in our Raspberry Pi Pico are much more limited. Here is a brief run down of data types, flow control, as well as creating functions.

Data types

Python supports quite a few different data types but despite the number of interesting data types it is more likely that the basic ones will be used in the programs that will fit onto the Pico.

bool	Boolean type that can be either True or False.
int	Integer type that can be a value from -2147483648 through 2147483647.
float	Float type represent machine-level floating point numbers. The values supported are dependent on the implementation for your platform.
str	String type is an immutable sequence of character values.
range(x,y)	This is an immutable sequence of numbers from x up until y.
list	Quite similar to what is normally considered an array except a list can contain different data types. ie. [3.14,'beginning of pi',false]

Flow control structures

Software development is the process of designing, creating and deploying solutions. Designing and creating is the process of taking your thoughts and ordering them to execute in the correct order.

Virtually all controlling functions can be created using an if/then statement and a goto[15].

15 https://homepages.cwi.nl/~storm/teaching/reader/Dijkstra68.pdf

Yet goto statements are a really ugly there are a lot of elegant structures that have been created since the goto was first discovered.

Perhaps the most important control structure in Python is making a comparison and branching based on that information.

If / then

```
testpct = 97
if testpct == 0:
    print("yikes!")
if testpct < 60:
    print("ouch!")
    grade = "F"
if testpct >= 60 and testpct< 70:
    grade = "D"
if testpct >= 70 and testpct < 80:
    grade = "C"
if testpct >= 80 and testpct < 90:
    grade = "B"
if testpct >= 90 and testpct < 95:
    grade = "A"
if testpct >= 95:
    print("wow!")
    grade = "A+"

print("your grade is ",grade)
```

If / elif / else

```
percent=88
response=""
message0="much work needed"
message1="ok, but try harder"
message2="very good"
message3="exceptional"

if (percent < 60):
        response=message0
elif (percent <80):
        response=message1
elif (percent <90):
        response=message2
```

```
else:
        response=message3

print("test score is ", percent, " percent");
print(response);
```

The while loop can be executed an unspecified number of times but it is not guaranteed to execute depending condition.

While loop

```
idx = 1
while idx < 10:
      print(idx)
      idx++
```

If a while loop fails to set a or update condition to cause it to exit it will continue forever. Another flow control structure which may offer a more defined run is the for loop.

For loop

```
for i in range(0,3):
      print(i)
```

This example will iterate the values from 0 up until 2 (ie 0, 1, 2). This loop will always run a fix number of times. Either of these values in the range can be replaced with a variable for more flexibility. Often loops are used for iterating through arrays but there is a more convenient way to processes arrays.

For loop

```
fruits = ["apple", "banana", "cherry"]
for x in fruits:
      print(x)
```

This second looping example will iterate all three times with the index variable being set to the each value in the array.

Yet, despite what every book or computer science teacher has ever said, we still have the goto statement. In Python it is now been disguised as *break* and *continue* keywords.

Break

```
endval = 100
for i in range(0,endval):
      complexvalue = I ** 2
    if complexvalue > 100:
          break
print i
```

This break example will calculate the variable i raised to the second power (ie i^2). If the calculated value is greater than 100 then the break command will cause the loop to exit prematurely, if the value i is examined for this example it will contain the value 11. This is essentially a goto end of loop.

However there are times that part of the loop should be circumvented but not exited.

Continue

```
total = 0
for i in range(0,10):
    if i % 2 != 0:
          continue
    print(i)
    total = total + i
print ("total",total)
```

The continue statement will skip the rest of the loop code beneath it and will resume control at the top of the loop. This is a goto top of loop. This continue example will sum up all of the even numbers between and including 0 through 9 resulting in a sum of 20.

The final, and perhaps most important, component in software development is to break up our solution into clear meaningful pieces which in Python is functions.

Functions

```
def calcarectangle(width,height):
        area = width * height
        return area

w = 10
h = 5
a = calcarectangle(w,h)
print("area of",w," x ", h," = ",a)
```

This function example shows a proper python program which will calculate the area of a rectangle and display the results.

MicroPython

A full install of Python is nearly 100mb and that is not suitable for a micro-controller. MicroPython is an implementation of Python that is mainly compatible with Python 3. It written in C and is small enough that it can run on a micro-controller.

The MicroPython "platform" compiles the Python code into bytecode as well as a runtime interpreter for that bytecode. In addition to normal python core libraries is a few additional modules that allow direct access to the hardware.

The MicroPython interpreter won't run on some of the original micro-controller boards such as the Arduino Uno but the Dual-core Arm Cortex-M0+ powering the latest generation of micro-controller boards have both the space for running the MicroPython interpreter as well as the processing power for an interpreted language.

Micropython REPL

When MicroPython is installed on the Pico it actually does a bit more than just run your programs. It comes with a built in read, evaluate, print loop interpreter which is referred to as the REPL. This loop allows you to enter statements to be evaluated. This is actually pretty cool as you can then try out some of your statements one at a time, create functions or even get a look at the contents of your variables.

In order to run python programs on the Pico, or run individual commands it is necessary to install MicroPython on the Pico. The first step is to download the interpreter and install it on the Pico. Simply execute the following command from a terminal window to download the most up to date MicroPython version.

```
$ wget
https://micropython.org/download/rp2-pico/rp2-pico-latest.uf2
```

The rp2-pico-latest.uf2 firmware is installed in the exact same way as installing a program on the pico. When connecting the Pico to your Raspberry Pi press the boot select button when attaching the USB cable. This will mount the Pico as a drive. Copying the previously downloaded micro-python file to the drive will cause it to install and disconnect the drive.

It is possible to connect to the Pico with a serial connection over the USB cable. With the Pico plugged in, you can use any terminal software to connect from your Raspberry Pi. Why connect to the Pico over a terminal? This is a convenient way to use the REPL interpreter running on the Pico.

One small terminal program that can be used is minicom. If minicom is not installed you can easily install it from the terminal with the following command.

```
$ sudo apt install minicom
```

Minicom is a command line program that runs in your terminal window.

```
$ minicom -o -D /dev/ttyACM0
```

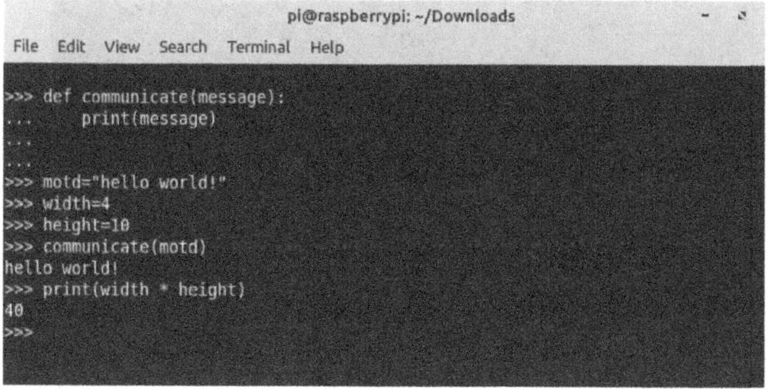

Figure 7: Using minicom to access REPL on Pico

In the MicroPython REPL you can enter any valid commands to be evaluated. This can be from simple variable assignment to entering your own functions, Figure 7.

However, the Pico is more than just a command line interpreter for simple text processing. The ability to control other devices is the strength but this is done by controlling the header pins.

Most of the interfacing between MicroPython and the 2040 chip is done using the machine module. This can be seen from our sample program entered the interpreter, Figure 7.

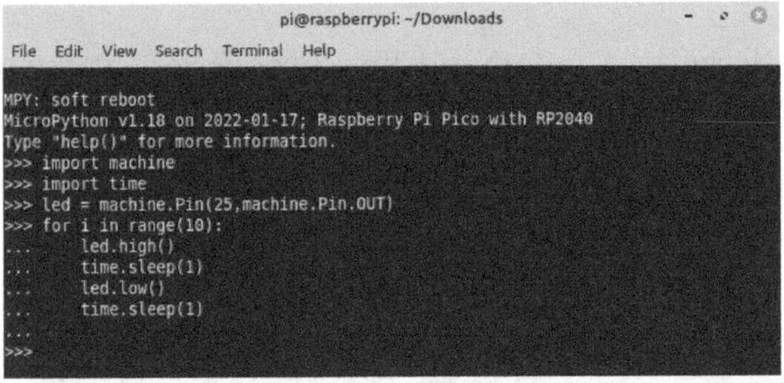

Figure 8: Python version of blink program

This example will flash the onboard led 10 times and then the REPL interpreter will resume.

Troubleshooting

It is possible to purchase a single Raspberry Pi Pico but sometimes there are better deals to be had if you purchase a couple. This is convenient if you think you have are having a hardware problem during development.

Having more than one Pico can also cause problems.

In my case the problem was with the minicom terminal emulator, it connected but did not appear to be working. That is, the REPL interpreter wasn't working. Eventually I found out the problem was that this Pico had not been prepared with the MicroPython. The solution was simple - attach the Pico as a drive and copy the MicroPython to it so it is installed.

5 THONNY DEVELOPMENT ENVIRONMENT

In the "olden" days developers wrote their program in a simple text editor. These editors allowed you to type in your program, save your program or perhaps load or debug your program. This was not the easiest way to write programs.

The solution to this rather manual approach was to create an integrated development environment (IDE). The "IDE" tends to be a combination of different tools. The most obvious is the editor itself. In a lot of cases the IDE can do much more than just type in text. The minimum functionality tends to be syntax highlighting support. This is the developers equivalent of spell checking which is built into your word processor. IDE's can help you become more productive but one of their major selling points is the ability to step through and debug your code.

There are a lot of other features your IDE might support, and one of the features might be that it is free.

The more powerful the IDE the more features it has and so the more setup it may require. Not only that, the IDE might not be free for non-personal use. Shucks it might not be free at all. Here are a few of the features you might find in your IDE.

Debugger
Syntax highlighting
Auto completion
Interpreter window
Variable watch window
Highlights syntax errors
Expression evaluator
Assistance for refactoring
Generates code based on templates
Tomcat server support
Integrated with source code central
Multi language support
Formatting source code
Plugin support

Thonny - Python IDE

One of the major advantages of the Raspberry Pi family of devices is the level of support that they receive. Some of this is due to support by a like minded community who enjoy the freedom these unique devices grant. Another form of support is that virtually all interesting software for the Raspberry Pi devices is open source. The software is both free in cost and free as in freedom to modify it as you please.

The "Thonny" Python IDE is not only free but

it is already installed on your Raspberry Pi. This helps to make the Raspberry Pi / Pico development combination a very smooth experience from setup through to use.

Thonny has all of the main features you would expect in an IDE with the most important being debugging in the IDE.

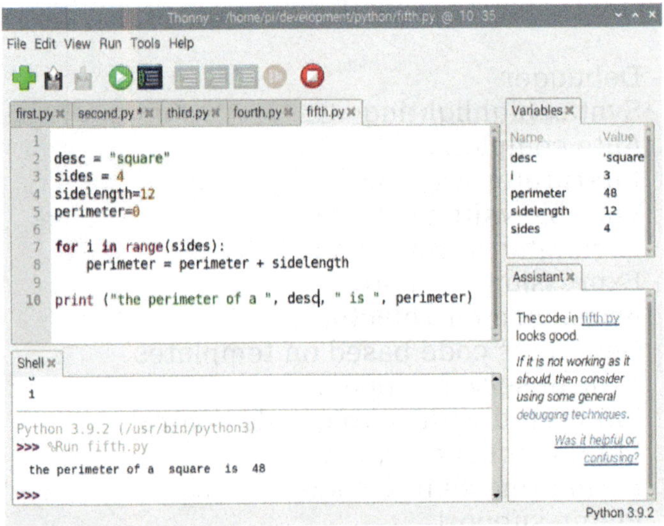

Figure 9: Thonny IDE running on Raspberry Pi

It is possible to run and debug by selecting menu choices from the Run menu but all of the most common choices have their own icon on the toolbar.

Run

This will run the current script in the IDE. The output will appear in the shell

		window.
	Debug	This icon will start the debugger at the first line. Once debugging in started you can use step over, step in, step out and resume buttons.
	Step Into	This will execute the current statement. If the statement is a function call, then the step icon will move into the function and execute the first statement.
	Step Over	Step over will execute the current statement. If the current statement is a function call, then the entire function call will be executed. The debugger will stop on the next line.
	Step Out	This will begin running and will continue to run until the current function is exited.
	Resume	This will continue to run until a break point has been hit.
	Stop	This will stop the debugger.

Setting debug break points is done by double clicking on the line number in the editor. This will create a small red dot to mark the location. Thonny can be used for development on your personal computer or it can also be used for developing on your Pico.

Initially it will be using the same interpreter as Thonny itself is using but it can changed to connect to the REPL interpreter on the Pico. Changing this is a simple matter of configuring a different interpreter in the IDE. This is done from the Run menu, Figure 10, you simply select "Select Interpreter ..." and then change the interpreter to be "MicroPython (Raspberry Pi Pico)".

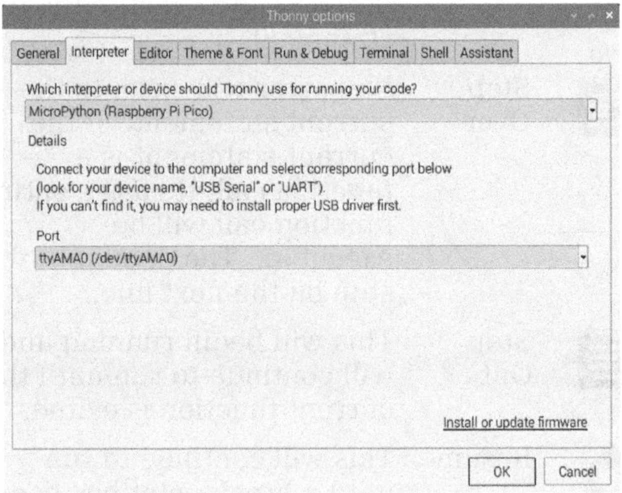

Figure 10: Setup an interpreter to run on the Pico

Once the interpreter has been changed you can type commands in the shell window and they will be executed on your Pico.

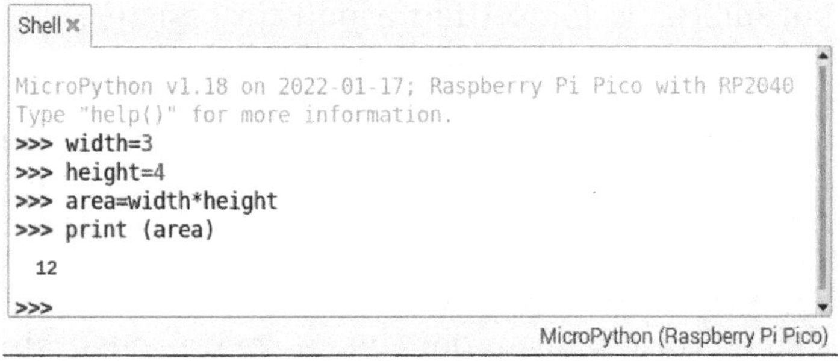

The Raspberry Pi Pico, just like many other single board controllers, has general purpose I/O pins and makes it easy for developers to use those pins. In order to keep everything organized, MicroPython has a number of different modules specifically for each platform. These modules provide the developer access the hardware functionality directly. More specifically the hardware has been encapsulated in a python module named "machine". The m0+ has two threads and it is possible to access the second processor core via the _thread module.

Controlling Pins - Pin class

Perhaps one of the most important classes is the Pin class. This class lets you control the pin as you see fit. This is quite convenient when you are directly controlling a device or component. The Pin class has an init function to setup all the attributes about the pin.

```
init(mode=- 1, pull=- 1,*, value=None, drive=0, alt=- 1)
```

The "mode" is for describing if the pin should be used for input or output. The "pull" parameter is for setting a pull up resistor, pull down resister or no resistor. Setting the output level of the pin can be done with the on(), off() as well as the toggle() method. Getting the value from the pin can be done with the value() function.

Simple input example

Using this information in a more concrete situation it is possible to use a button to get input from a user.

```
import time
from machine import Pin

button = Pin(28, Pin.IN,Pin.PULL_UP)
led = Pin(15, Pin.OUT)

def buttontest():
    cnt = 0
    runseconds = 10
    for i in range(runseconds * 10):
        if button.value() == 0:
            led.toggle()

        if cnt % 10 == 0:
            print ("second " ,cnt/10+1)
        time.sleep_ms(100)
        cnt = cnt + 1

buttontest()
```

Listing 2 Reading button input

The buttontest() method has a loop that runs over 10 seconds. Every 1/10th of a second it will sleep for 100 milliseconds but when it is not sleeping it will do a test if the button has been pressed.

The button variable definition is setup for input (Pin.IN) and more importantly it has a pull-up resistor (Pin.PULL_UP).

```
button = Pin(28, Pin.IN,Pin.PULL_UP)
```

Because of the pull-up resistor the value retrieve method will return a 1. When the button is pressed the button will connect the pin GP28 with ground.

When the pin GP28 is connected to ground the value method will return a zero.

Figure 11: Pico with a button and a led

It might be difficult to see the wiring on Figure 1, there are essentially three separate setups.

Ground setup
It is easy to overlook pin 23(ground). The ground pin is connected to the ground line on the breadboard. This makes it convenient to connect any parts directly to ground without stretching wires across circuit.

Led setup
Pin 20(GP15) connects to the anode of the led. The cathode of the led is connected to a resistor and the resistor is connected to ground line on the breadboard.

Button setup

Pin 34(GP28) is connected to one leg of the button. The second leg of the button is connected to ground line on the breadboard.

It might appear that there are four pins but they are actually pairs (internally connected.

Complex button example
The previous simple button example clarifies the technique for reading from a button. However there are not so many situations where a single button will provide all the functionality necessary. What might happen more often than not would be some form of input device such as a keypad.

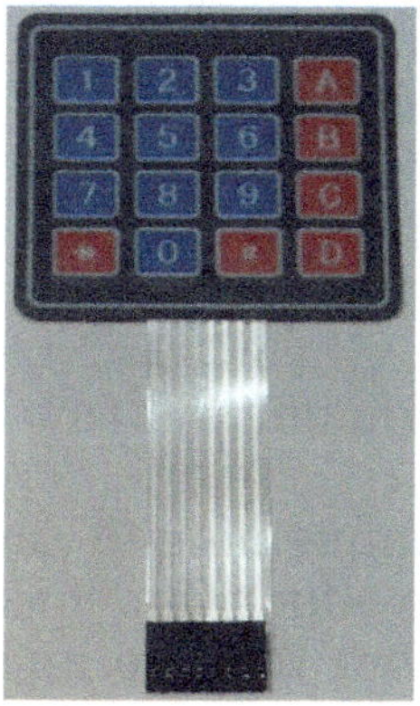

This is a 4x4 Matrix Array Keypad Keyboard which can be used for entering simple passcodes. This device is cleverly designed so it can be controlled by eight connections instead of 16 which could be required in a very naive keypad implementation using two pins per button.

The keypad is indeed a matrix where all keys on the same row are connected together and all keys on the same column are connected together. This particular setup has the advantage of only requiring eight connections but the disadvantage of only allowing a single key to be pressed at a time. This might not sound like a limitation but standard computer keyboards can allow up to six keys to be simultaneously pressed.

Each of the buttons are in the open position so only by pressing the button will complete the circuit. This keypad works by passing a signal on the row and testing the column to see if the signal comes through. If the signal does come through we will know which row we passed the signal and which column we tested thus we know which key on our keypad was pressed.

The algorithm for testing for a key-press is as follows.

1. Loop through all the rows
2. Loop through all the columns
3. Pass current through current row
4. Test current column if current detected
5. If detected, lookup the key name
6. pause
7. discontinue current through current row

This algorithm has been implemented in Listing 3 and each step in the algorithm has been implemented in our code but the one line of code that is not self explanatory is step 6, do a pause.

```
for rowidx in range(0,4):
    for colidx in range(0,4):
        row_pins[rowidx].high()
        if column_pins[colidx].value() == 1:
            retkey = keyarray[rowidx][colidx]
            time.sleep_ms(500)
        row_pins[rowidx].low()
```

Listing 3 code to test for keypress

The pause is to give the user enough time to remove his/her finger from the keypad. Without this pause the program will immediately advance to the next row and will pass the current through. Because the key, which has not yet been released, will seen as selected but on the next row. This could be make it appear that many keys have been pressed.

The logic for collecting an entire pin code is not terribly difficult once you can capture a single button press. Simply loop collecting key presses until the exit condition signals the code is complete. The algorithm for using this keypad get and test a passcode.

```
Disable access
loop until successful code
    get a password attempt
        reset key presses
        loop until special char
            get single keypresses
        return keypresses
    compare attempt against secret passcode
    if attempt is correct
        then enable access
```

Most of the logic for using this keypad for a security door actually is the collection of the passcode. A sample of what that code would look like can be seen in Listing 4.

```
def getline():
    done = False
    keybuffer = []
    val = None
    while done == False:
        val = getkey()
        if val == '#':
            done = True
        else:
            keybuffer.append(val)

    # return as a string
    retval = ""
    for single in keybuffer:
        retval += single

    print("retval line = " + retval)
    return retval
```

Listing 4 Collect a sequence of key presses

The eagle eyed reader might notice that there are some print statements in this function. These print statements were used for debugging during the development in the Thonny IDE. If these print statements were left in the code it would have no impact in a real world project but it would also not help outside of the development environment.

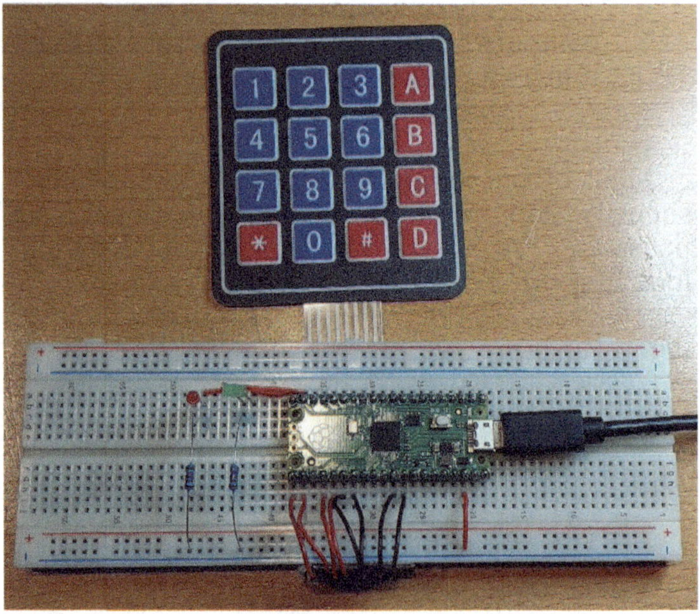

Figure 12: Pico powered "door unlock keypad" setup

This solution has red and green leds to provide feedback on the state of the “door”. The red led is lit until the password has been successfully entered. Once that happens, the green led is lit. If this were an ideal solution it would beep with each key press to provide some additional feedback to the user.

The code for controlling the keypad is listed in Appendix - keypad example.

Simple output example

Because a micro-controller board doesn't typically come with a screen, led's are a common way of providing user feedback. The simplest possible case is to simply turn on/off a led.

Listing 5 will blink a led five times with a ½ second on and ½ second off. This program will cause the onboard led to blink however the logic is identical for using a different GPIO pin to control an external led.

```
import time
from machine import Pin

LedPin = Pin(25, Pin.OUT)
for i in range(10):
      LedPin.off()
      time.sleep_ms(500)
      LedPin.on()
      time.sleep_ms(500)
```

Listing 5 Blink onboard led

You can create a project that has a LED which turns on, you can limit how bright that led is by adding a resistor.

Depending on which resistor is used will create differing levels of brightness. This is problematic if you want to control the brightness level from your program. This can be controlled from software by using pulse width modulation (PWM).

Pulse width modulation is the technique to switch the LED on and off quickly. When a LED is on 50% of the time it appears to the eye that the LED is approximately half as bright when compared to a LED that is on 100% of the time. This process can be programmed manually but this technique is so common that it has been implemented in hardware.

In order to use PWM you need to know a few terms.

Duty Cycle
Frequency

The Duty cycle represents the amount of time that the signal is high. When signal is high half of the time and low half of the time then the duty cycle is 50%. When the signal is high 30 milliseconds and low 70 milliseconds then duty cycle is 30%.

$$duty\ cycle = \frac{on\,time}{total\,time} \qquad duty\ cycle = \frac{30\,ms}{100\,ms}$$

The frequency is the number of complete cycles that the signal goes high and low.

$$Freq = \frac{1}{on + off} \qquad Freq = \frac{1}{30\,ms + 70\,ms} = 10\,\%$$

```
from machine import Pin, PWM

# create PWM object from a pin
pwm0 = PWM(Pin(0))
print("our PWM object")
print(pwm0)

# get current frequency
print("current frequency")
print(pwm0.freq())

# set frequency to 1000ms
print("setting frequency")
pwm0.freq(1000)
print(pwm0.freq())

# get current duty cycle
print("current duty cycle")
print(pwm0.duty_u16())

# set current duty cycle
print("setting the duty cycle")
pwm0.duty_u16(200)

# turn off PWM on the pin
pwm0.deinit()
```

Listing 6 Pulse width modulation calls

Simple timing example

With single board computers and micro-controllers there are a lot of situations were timing is very important. The good news is that the Pico has super fine granularity which makes it possible to have delays in the microsecond range but also for the milliseconds or seconds, Listing 4.

```
import time

# sleep for 1 second
time.sleep(1)

# sleep 1000 milliseconds (1 second)
time.sleep_ms(1000)

# sleep 1000 microseconds (1 millisecond)
time.sleep_us(1000)

# get millisecond counter
start = time.ticks_ms()
time.sleep_ms(23)
end = time.ticks_ms()

# compute time difference
delta = time.ticks_diff(end, start)
```

Listing 7 Example showing time arithmetic

This precision timing comes is convenient when doing communication. This timing will be used by the Pico when doing I2C communication.

First Program

The Raspberry Pi Pico is almost as powerful as the first generation iPhone. This makes it more than capable for doing calculations. The simplest "first program", well after the traditional typical hello world, would be to do a few math calculations. Digging deep into my algebra book are the formulas for calculating both the area and the circumference of a circle.

$$area = \Pi \; x \; radius^2$$

$$circumference = 2 \; x \; \Pi \; x \; radius$$

The first time the program is run it will need to be saved either locally or to the Pico as Thonny will save the program prior to each run.

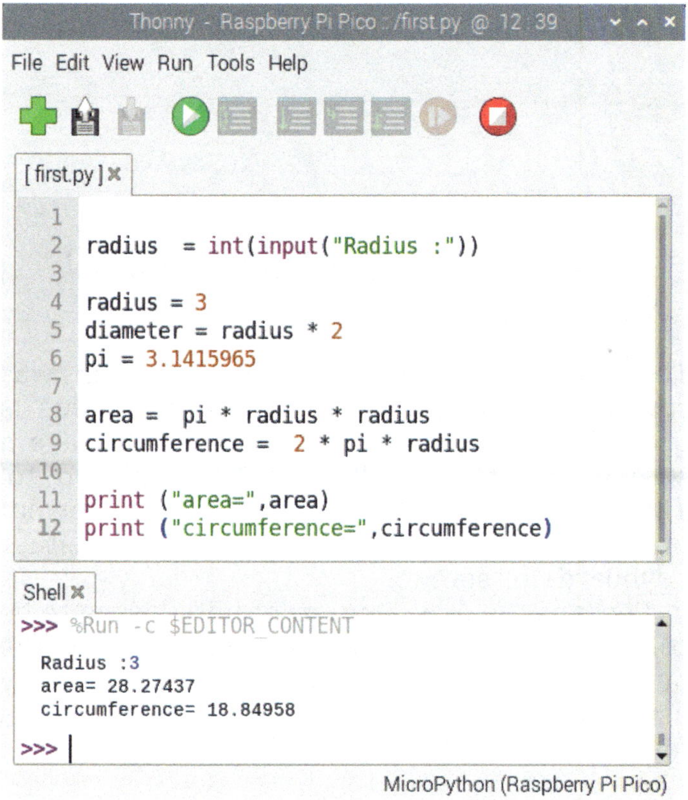

When this program is run you will be prompted to enter the radius. This is only possible when run on the host machine. The "input" function will wait until the user enters the value and then the program will continue.

I.e.

```
variable = input("prompt text")
```

6 PICO AND SCREEN OUTPUT

The last task in the previous chapter was running a small calculation program. This is great for traditional computers as they have a screen that the results can be displayed to. This program specifically is somewhat problematic because the Pico has neither keyboard nor screen.

Perhaps the most important initial issue to be solved is the ability to communicate with the outside world. One possible way to do this is by using a rather tiny LCD screen. There are a few different types to choose from.

20 x 4
16 x 2
16 x 4

This is by no means an extensive list, there are a lot of interesting small LCD screens that can be used not forgetting that e-paper screens also exist in quite a few different sizes.

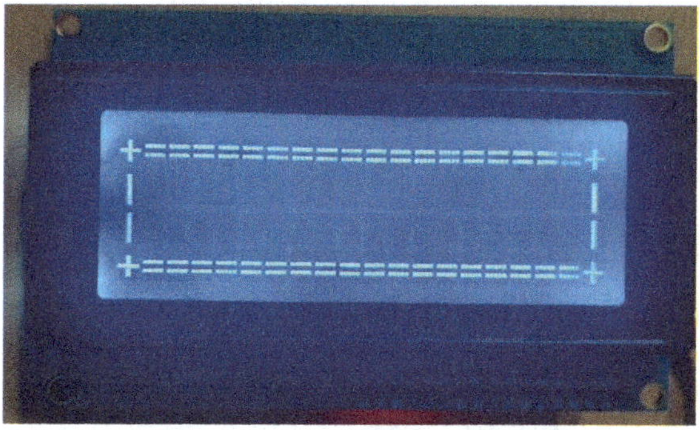

Figure 13: 20x4 character LCD display

To get started I will be using a small I2C screen that is four lines each with 20 characters, Figure 13. This screen was left over from another micro-controller project written in C to be used with the Raspberry Pi.

I2C Communication

Using an I2C bus requires very specific timing watching the data line (SDA) for address bits, ack/nack bits, read/write bit and data bits. All of this is while verifying if the signal is high or low based on the clock line (SCL).

It is possible to write your own low level routines to read and write the bits onto the I2C bus, although this is tedious. Creating your own routines would be error prone and would not add much in the way of value to a solution. The good news is that this is not necessary as it has already been done for us in the I2C classes. Two different classes have been provided - SoftI2C and I2C.

The former is a purely software solution and can be used on any output capable pins. This is implemented by "bit banging". This is just a fancy way of saying software drives the solution without any hardware support. The I2C class is functionally the same as the SoftI2C class.

Constructor	Define the sda pin, scl pin and the frequency.
scan	Returns an array of I2C addresses
readfrom	Reads a number of bytes from a given address.
writeto	Writes a number of bytes to a given address.

Writing your own I2C support routines is challenging but using these routines makes the process almost trivial.

I2C LCD Screen

The internet is full of talented people who also have interest in using different types of hardware in their projects and are willing to share their solutions.

On Github there is a project[16] by Dave Hylands which supports not only this screen for the Pico but can also support different sized screens for a number of different embedded platforms. You can download the entire project which consists of a number of different source files to support micro-controller boards. In order to use this screen, only two files were needed from this project.

machine_i2c_lcd.py

16 https://github.com/dhylands

lcd_api.py

Although the screen is a I2C device, these routines will make it painless to use. However, in order for our program to use them they will need to be saved onto the Pico.

To transfer them to the Pico simply open these files in the Thonny IDE and when doing a saveAs you are prompted where they should be saved. If you pick the "This computer" a dialog will be opened allowing you to find a location on your host, but if you choose "Raspberry Pi Pico" the file will be saved onto the Pico. All of the files that are saved on the Pico will be[17] displayed on the IDE on the left side, Figure 14.

17 Well, the files can be displayed depending if you select the files to be viewed from the view menu.

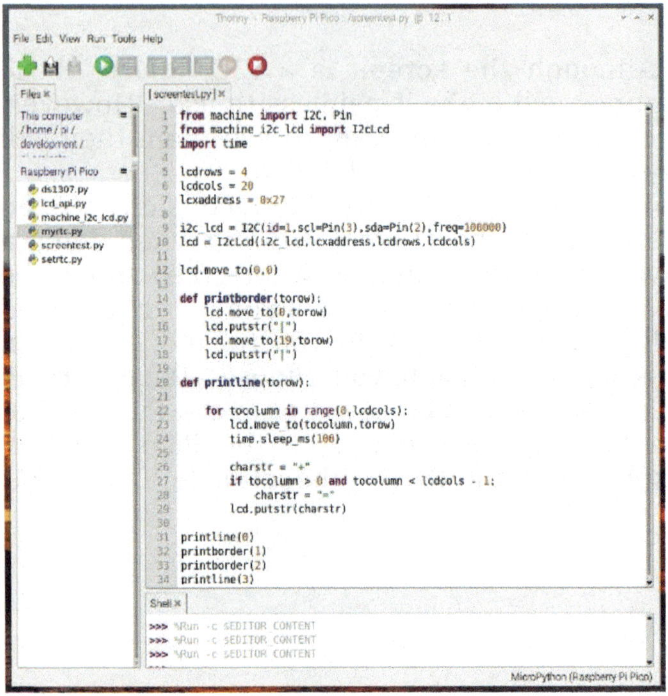

Figure 14: Thonny IDE, the window on the left shows files saved on Pico

The program visible in the Thonny IDE, Figure 14, will print a small text frame on the 20x4 LCD screen, Figure 13. This code can be decomposed into four different major sections.

includes
global variables
functions to draw
main code

Although this code is fairly explanatory, each of these sections of Listing 8 will be described.

```
1  from machine import I2C, Pin
2  from machine_i2c_lcd import I2cLcd
3  import time
4
5  lcdrows = 4
6  lcdcols = 20
7  lcdaddress = 0x27
8
9  i2c_lcd=I2C(id=1,scl=Pin(3),sda=Pin(2),freq=100000)
10 lcd = I2cLcd(i2c_lcd, lcdaddress,lcdrows, lcdcols)
11
12 def printborder(torow):
13     lcd.move_to(0,torow)
14     lcd.putstr("|")
15     lcd.move_to(19,torow)
16     lcd.putstr("|")
17
18 def printline(torow):
19
20     for tocol in range(0,lcdcols):
21         lcd.move_to(tocol,torow)
22         time.sleep_ms(100)
23
24         charstr = "+"
25         if tocol > 0 and tocol < lcdcols - 1:
26             charstr = "="
27         lcd.putstr(charstr)
28
29 printline(0)
30 printborder(1)
31 printborder(2)
32 printline(3)
```

Listing 8, print boarder around LCD screen

Imports, lines 1-3

This program includes three different sets of includes. The first is importing the I2C class and Pin class from the machine module. The machine module is a special module that has all of the hardware specific code for our device. The methods in this module may vary slightly when using different micro-controller boards.

You need to look quite closely at the second import line. This will import the I2cLcd class from the "machine_i2c_lcd" module; This module is one of the two files we downloaded from gitlab[18].

The third and final include is time. This is a common micro-python module supporting the time methods.

The idea that some routines need to be included is not foreign to most developers but seeing these two different styles of import is perhaps a bit jarring. The difference is that the "import name" will import the entire module while the "from module name import object" will only import a specific member from that module.

Global variables, lines 5-10

The screen is physically four lines of twenty characters so it should come as no surprise that lcdrows=4 and lcdcols=20 exist but what does lcdaddress=0x27 represent?

This LDC screen is an I2C device and like all I2C devices it has an address which must be used to identify that device out of the many that could be connected to your I2C bus. The address of this device is 0x27 hexadecimal or 39 decimal.

This address is being used with the Pico specific I2C class along with some of the other

18 https://github.com/dhylands

important values required for I2C setup.

```
i2c_lcd = I2C(id=1,scl=Pin(3),sda=Pin(2),freq=100000)
```

The i2C_lcd variable is actually used for dealing with the I2C logic. This variable is an input into our actual variable for controlling the LCD screen.

Parameters explained

id=1	Which of the two I2C buses you are connected to (0 or 1 see pin layout)
scl=Pin(3)	Scl takes a Pin object which supports the system clock (see pin layout)
sda=Pin(2)	Takes a Pin object which supports data line (see pin layout)
freq=100000	supported frequencies (100k,400k,1000k)

Note:
It is not necessary to pass in the id value. When this value is not provided the I2C will be performed using software instead of using the hardware support.

```
lcd = I2cLcd(i2c_lcd,lcdaddress,lcdrows,lcdcols)
```

The lcd variable actually controls where on the screen the output will be sent. This variable, I2cLcd object, knows nothing specifically about how to communicate using

the I2C protocol but it will use the i2c_lcd variable which only knows how to do that.
Parameters explained

i2c_ldc	The I2C object.
lcdaddress	The address of the I2C LCD device
lcdrows	Number of rows supported by LCD
lcdcols	Number of columns supported by LCD

Functions, lines 12-27
These two functions will output the necessary parts of the text border to the LCD. The printborder method will put a vertical line at the beginning (far left side) and the ending (far right side) of the line. The printline function will output the entire line to the given row.

Main code, lines 29-32
The whole program comes down to these four lines. The functions will use some of the global variables such as the screen and screen size and these four lines will call the functions in the proper order.

Diagram
Connecting the LCD to the Pico is not difficult. There are only four pins on the LCD that need to be connected

SDA

SCL
VCC
GND

The connections are fairly simple, the SDA and SCL need to be connected to a similarly named pin on the Pico. There are two I2C controllers that can be used with many different pins. Which of the two controller you choose is unimportant for the functionality but is more connected to the layout of your device. Some pins might be better located for wiring up depending on your circuit layout.

I2C Controller	GPIO Pins
I2C0 - SDA	GP0/GP4/GP8/GP12/GP16/ GP20
I2C0 - SCL	GP1/GP5/GP9/GP13/GP17/ GP21
I2C1 - SDA	GP2/GP6/GP10/GP14/ GP18/GP26
I2C1 - SCL	GP3/GP7/GP11/GP15/ GP19/GP27

Telling the time

Different projects may require different types of hardware. This might be simple button, leds and resistors or it could be a more complex device that uses I2C for communication. In some cases it might even be necessary to use multiple I2C devices. One such example, although perhaps a bit contrived, would be creating a digital clock.

The Pico does have the ability for very precise timing down to the microsecond level but this is not quite the same thing as a real time clock calendar module.

These real time clock modules (RTC) have been available for quite a few years to support older micro-controller platforms. It is possible to purchase one with a battery backup on the internet, perhaps even cheaper than the battery itself.

Communication between most of these modules and their host (Pi, Pico, Arduino,...) is done via the I2C protocol. This works out well, as we can test a Pico project with multiple I2C devices. The task is pretty strait forward. Read the time and update the screen with the new time.

Quite a bit of the code for this example, Listing 9, is similar to the previous example. It is not terribly difficult dealing with the low level parameter passing to I2C devices but it can be a bit fiddly which is why I am using the ds1307 module[19] from gitlab.

```
from machine import I2C, Pin
from ds1307 import DS1307
from machine_i2c_lcd import I2cLcd
import time

lcdrows = 4
lcdcols = 20
i2c_lcd =
I2C(id=1,scl=Pin(3),sda=Pin(2),freq=100000)
lcd = I2cLcd(i2c_lcd,0x27,lcdrows,lcdcols)


```

19 https://github.com/mcauser/micropython-tinyrtc-i2c

```
   i2c_rtc =
13 I2C(0,scl=Pin(1),sda=Pin(0),freq=400000)
14 rtc = DS1307(i2c_rtc)
15
16 while True:
       time.sleep(1)
17
18     (Y,M,D,day,h,m,s,ms)=rtc.datetime()
19     timestr = "Time: %02d:%02d:%02d" % (h,m,s)
20     datestr = "Date: %4d-%02d-%02d" % (Y,M,D)
21
22     lcd.move_to(2,1)
23     lcd.putstr(timestr)
24     lcd.move_to(2,2)
       lcd.putstr(datestr)
```

Listing 9, program to read the time and display it on the LCD screen

This real time support module, just like the LCD module will also need to be copied over to the Pico so it can be used by our program. This is included on line 2 and our rtc object is created on lines 12 and 13. Retrieving the time and date is a simple call, line 17, which will return the time and date.

This program works fine but it does of course expect that the RTC has been previously set with an actual date and time. This can be done with the following code. I did this as a separate and independent step.

```
now = (year,month,date,day,
          hour,minute,second,0)
rtc.datetime(now)
```

Troubleshooting

The Thonny IDE is a great tool but like any development environment sometimes problems occur and the most important thing is how to recover.

While working on an unrelated piece of code, I somehow created a situation where my IDE no longer seemed to properly work with the Pico. It appeared that the Pico was somehow locked up or not responding. It was no longer possible to load or save files on it nor could commands be run on the Pico from the shell. It appeared that the Pico was bricked.

In this case, bricked could simply mean poorly programmed. If you save your program on Pico with the name main.py then that is specific program will be run on power up. This will prevent you from using the interpreter. If you find yourself in a similar situation, here are a few ways to correct the situation.

Delete main.py from Pico
Reinstall MicroPython to Pico
Nuke the Pico's contents
Reinstall Thonny
Reinstall Rasbian

This list is ordered by level of destruction. The first three on this list is dealing with a problem on the Pico itself. The "Nuke" option is the most serious. This will wipe out all bits and bytes on the Pico. The Pico will then be essentially at the same state as when it was first received. Any files that were saved on the Pico will be lost. The Nuke[20] process is simply uploading a special image in the same way any program is uploaded to the Pico. The first three

[20] https://www.raspberrypi.com/documentation/microcontrollers/raspberry-pi-pico.html#resetting-flash-memory

solutions deal with the Pico problem but there is another possible problem.

No software is perfect and it is possible to have problems not only on your client but also on your host. I encountered one specific problem that made it seem that I had bricked the Pico - I had not. In the end, the problem appeared to be on the host environment. The Pico worked fine, the solution was a fresh Raspberry Pi OS installation.

Before proceeding with these steps you should see if the problem is with your host setup. This can easily be checked by running the minicom terminal software from the host to the client. If the REPL loop is there and works then the problem is most likely ***not*** *the Pico.*

7 SMILING HARDWARE

The tricky part of learning a new programming language or developer task is usually finding a problem that needs to be solved. In this case the task needs to involve a micro-controller. Perhaps rather than solving a very complicated problem as our first project it would be better to create a fun little toy.

What I have always wanted was to create some sort of electronic pumpkin which I could put into the windows leading up to Halloween.

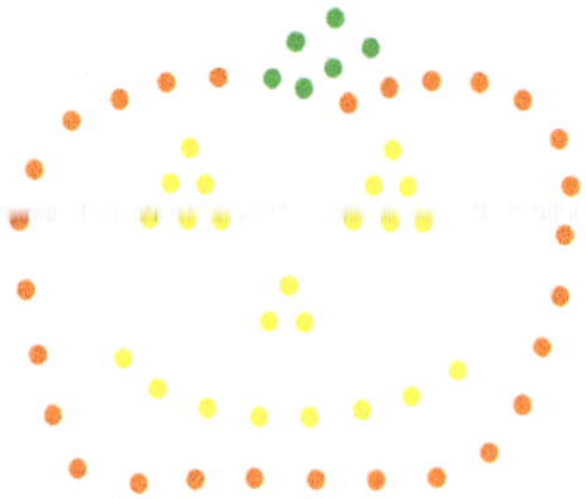

This could be quite a big task to try and

control 56 leds without some fairly tricky electronics. There is a technique for controlling more leds than the number of actual control pins that you have - this technique is called charlie plexing. This pumpkin face could be controlled with 8 pins but it is a very complex setup. Perhaps an easier task would be a small smiling face, Figure 13, with some form of animation.

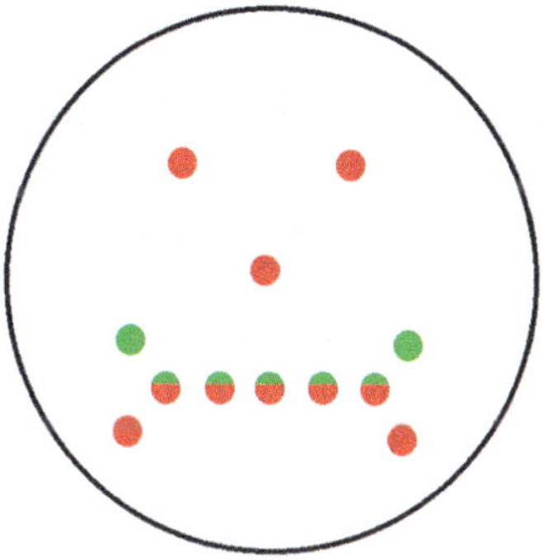

Figure 15: Small face with animation

This example "toy" could be made to smile or frown by simply controlling four leds at the edge of the mouth. Lighting up the top two makes a smile while lighting the bottom two makes a frown. The rest of the leds could just as easily be constantly lit with no other controls. Thus the change from happy to unhappy could be controlled with four pins from the Pico with all the rest being hard wired to on.

Controlling one led

A simple manual circuit to control a led might look somewhat similar to the following.

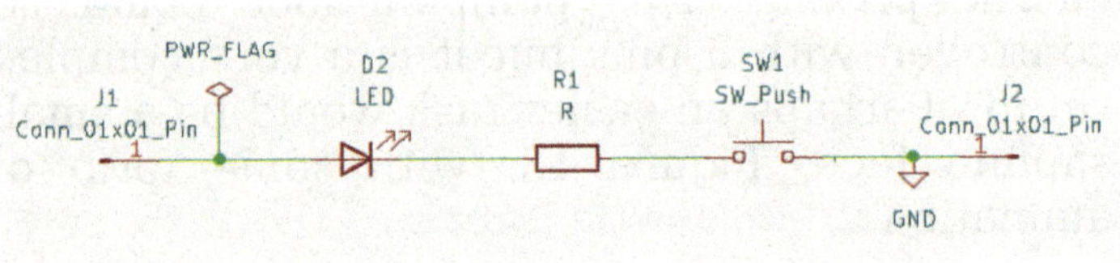

Figure 16: Circuit with manual button

Pressing button SW1 from this circuit will turn on the led as long as it is pressed. We can get a similar behavior by using one of the Pico's GPIO pins as the switch. Using the Pico allows our program to control the "on" state of the led. This is done by the Pico completing the circuit by outputting the power to the led. The power flows through the led to ground completing the circuit.

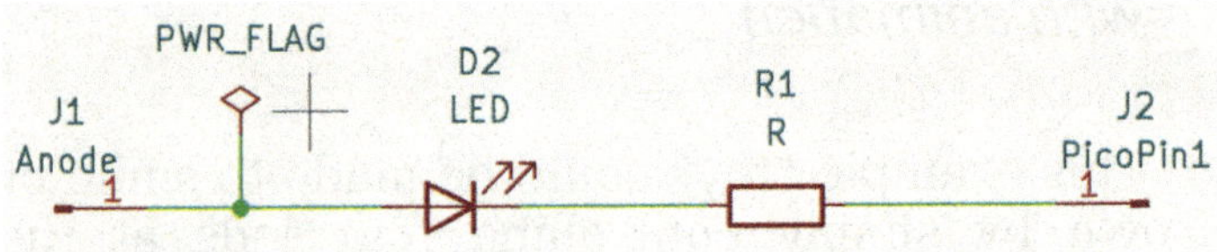

Figure 17: Circuit fully controlled by Pico

Smiley toy V1

Creating the smiley face toy prototype that is controlled by the Pico is just as simple as pushing a few led's through a piece of cardboard.

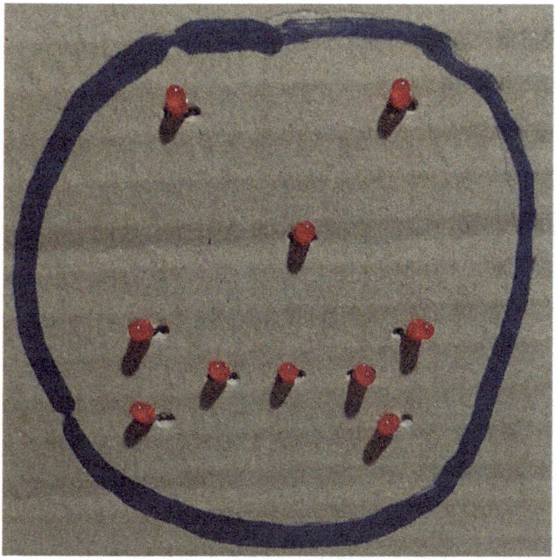

Figure 18: Smiley face prototype

What would be easiest if all of the leds that don't change (ie eyes, nose and part of mouth) were controlled by a single GPIO pin. That would be ideal but it is not possible because the Pico is not a power-plant and can only provide a limited amount of power. To get round this limitation I have wired up each led to a different GPIO pin.

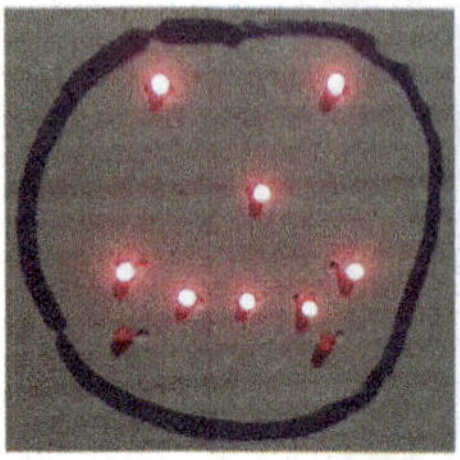
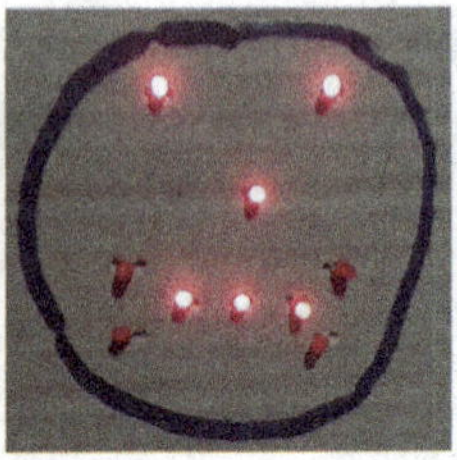
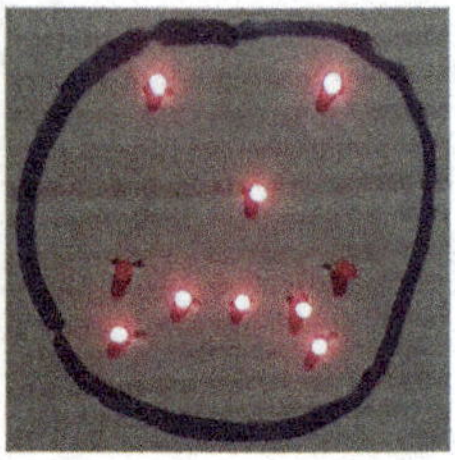

Smile Neutral Frown

The smiley doesn't have a lot of complicated logic so we can break up each expression into a couple of functions. Each of these facial expressions should be in their own method. In addition to these we should have one additional method to turn off all the leds. A method for turning off all of the leds is not strictly required but it is easier to turn of all then to keep track of what is on and then to turn them off.

The code is self explanatory. In the beginning a few variables are defined with each one controlling a single GPIO pin each of which controls single led. When this script it is run it will cycle between smiling and frowning.

```
import machine
import time

lefteye = machine.Pin(6, machine.Pin.OUT)
righteye = machine.Pin(7, machine.Pin.OUT)
nose = machine.Pin(8, machine.Pin.OUT)
smile1up = machine.Pin(9, machine.Pin.OUT)
smile1dn = machine.Pin(10, machine.Pin.OUT)
smile2 = machine.Pin(11, machine.Pin.OUT)
smile3 = machine.Pin(12, machine.Pin.OUT)
smile4 = machine.Pin(13, machine.Pin.OUT)
smile5up = machine.Pin(14, machine.Pin.OUT)
```

```
smile5dn = machine.Pin(15, machine.Pin.OUT)

def init():
    print("init")
    lefteye.low()
    righteye.low()
    nose.low()
    smile1up.low()
    smile1dn.low()
    smile2.low()
    smile3.low()
    smile4.low()
    smile5up.low()
    smile5dn.low()

def neutral():
    print("neutral")
    lefteye.high()
    righteye.high()
    nose.high()
    smile2.high()
    smile3.high()
    smile4.high()
    smile5up.low()
    smile5dn.low()
    smile1up.low()
    smile1dn.low()

def smile():
    print("smile")
    neutral()
    smile1up.high()
    smile5up.high()

def frown():
    print("frown")
    neutral()
    smile1dn.high()
    smile5dn.high()

def mainbody():
    init()
    for i in range( 11):
        smile()
```

```
        time.sleep(2)
        neutral()
        time.sleep(2)
        frown()
        time.sleep(2)
        neutral()
        time.sleep(2)
        init()

mainbody()
```
Listing 10 Smiley version 1 code

Once the script is fully debugged[21] it can be modified in a couple of ways. The first would be to modify the main function "mainbody" to be written as an endless loop. This will cause our face changing to continue forever. A second and more important change would be to save the file as main.py on the Pico. When a Pico is powered on it will automatically run a script called main.py.

[21] The print statements were used during development but would offer no functionality when it is "just powered on".

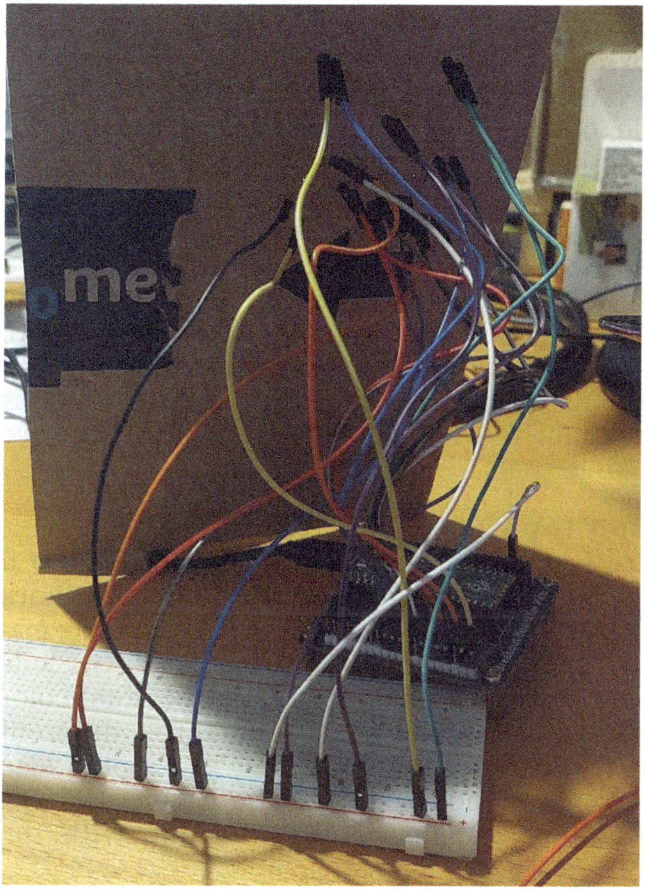

Version 1 of the smiley face is really quite simple although a bit messy when wired up. This would have been even more messy if we were controlling more leds.

If the number of leds is increased we would have a higher resolution face but as the number of leds increases so does the challenge in correctly wiring it up or powering it.

A different method of creating a more

interesting little toy would be to add some color. Using RGB leds would be ideal as they allow us to change the color of the eyes, the smile or the frown. It would be possible to cycle through the different facial expressions while also cycling though a large color palette at the same time. This would make for a really interesting project but it would require quite a few control pins as RGB leds have three cathodes and one anode (or vice versa). Changing this small smiley face v1 with only 12 RGB leds would then have 36 cathodes which is more than the 26 multi-function pins that are available to us.

Smiley toy V2

There is a middle ground that we could use if we still want to make our smiley face more interesting without going full RGB. If normal single color leds are too boring and RGB leds are too complicated we could opt for dual color leds. Dual color leds can be three different states, color 1, color 2 or off. This would also require 24 cathode pins for controlling a 12 led face, which should be easy enough with the Pico.

Controlling one dual color led

A simple manual circuit to control the different colors of the led might look somewhat similar to the following.

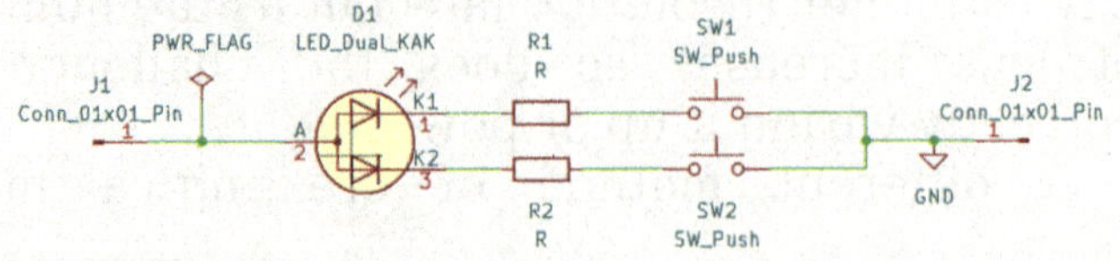

Pressing button SW1 will enable the first color while pressing SW2 will cause the second color to turn on. Using the Pico in this situation simply replaces both buttons with two of the control pins to complete the circuit.

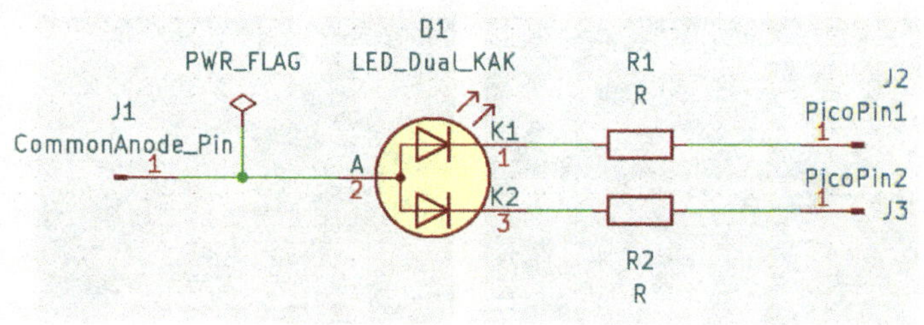

The circuit diagram for the "smiley" toy would connect each of the leds cathode's to a Pico pin.

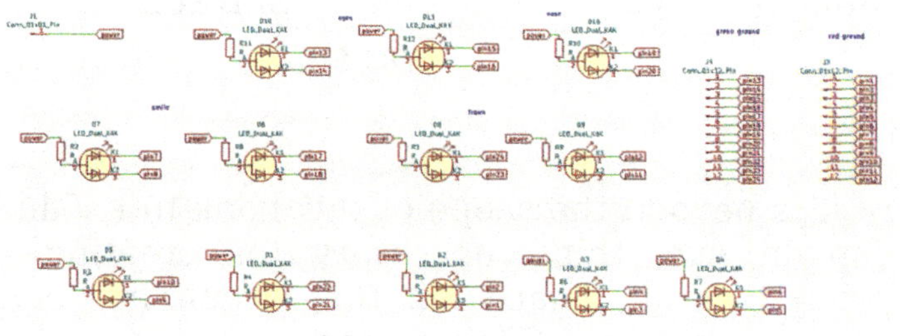

Figure 19: Smiley toy schematic

This circuit schematic is a bit odd at first glance as there appears to be no connections. Each of the small labels, pin1, represents a different electrical connection. The PCB development software keeps track of which led pins connect through to the Pico GPIO pin. Indeed, these are not the best descriptions for labels. It might make more sense if the label was the name of the Pico pin (ie GP7), or perhaps if the label was a meaningful name for the circuit (ie. LEFTEYE).

The circuit board for Figure 19 can be seen in Figure 20.

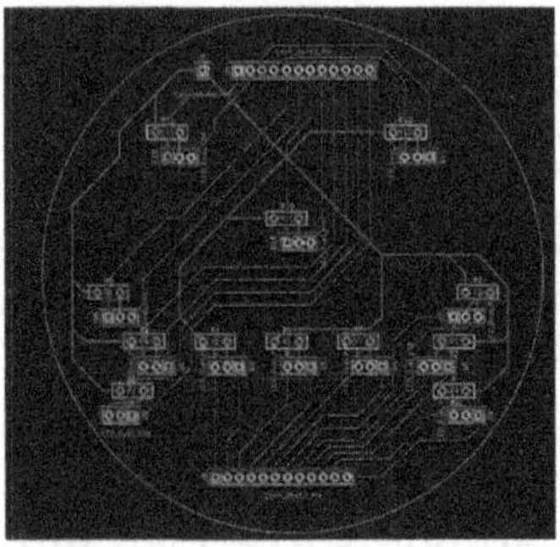

Figure 20: PCB for smiley toy v2

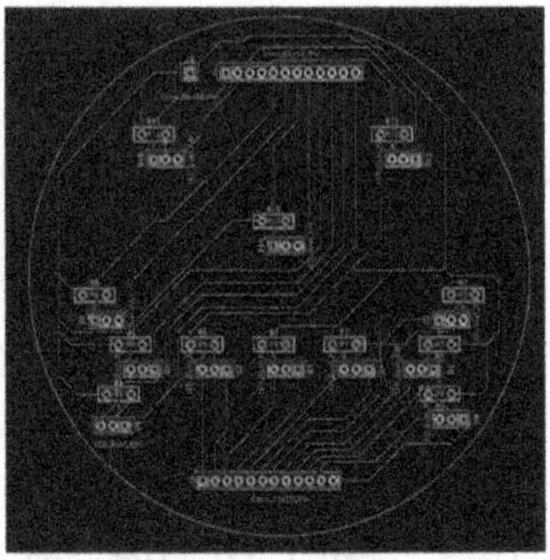

Figure 21: PCB all electrical traces on one side

It is beyond the scope of this book to try and explain everything necessary for creating a printed circuit board but this circuit board is composed of four different parts.

Resistor
Led
12 pin connector
1 pin connector (power)

The resistors are to ensure that not too much current is passed through each of the leds. These 12 header pin connectors were put there as a convenient way to gather all of the connections to two locations on the board. Ideally this could have been using a ribbon cable which would then be attached to the Raspberry Pi with as a single cable.

The lines on the circuit board, Figure 20, are the connections between all of the components. It might be difficult to recognize that most of these connections are red in color and a very few are light blue color. These actually represent two things, the first is that each of these colors represent that the electrical trace is on a different side of the PCB, the red traces are the front side of the board while the blue are on the back.

Another more important lesson that I should have remembered is that you should never rush[22] your PCB designs. After this was designed and ordered I spent a bit more time on my design and was able to route all of my electrical traces on one side of the board[23], Figure 21.

Overkill?

Is creating a circuit board actually overkill for this sample smiley project. Yes this is overkill but often difficulties arise due to bad connections. In addition it should be remembered that even though this smiley PCB

22 Not rushing PCB design, or importantly ordering a PCB. Unlike software which can be easily re-released, hardware is less malleable once it has been ordered.

23 If you are getting your PCB's manufactured single sided may be cheaper.

has 12 dual color leds and it would require 36 wires any of which could be a connection problem. In addition to simplifying the connections the PCB also provides form to the face so the eyes or mouth doesn't sag.

This alone is a good reason for creating this PCB. Having a PCB helps but it is still difficult to create a working electronic toy even with a PCB and our pile of parts. This difficulty is best shown by my first ever attempt to create a smart 7 segment display with controller. It actually didn't work because so many wires were packed into a small spot and I had a connection problem. This tends to not be a problem when using your own circuit board.

First self designed – front

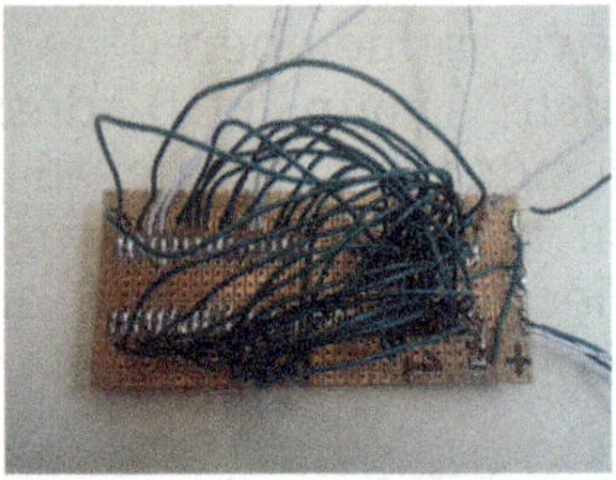

First self designed – back

The program v2

The code to control version 2 of the smiley face is actually not substantially different than version 1 described in Listing 10 Smiley version 1 code. One slight difference is that now there are two colors which need to be switched on and off. I have put all this information into a table to make it easier when creating the actual code. Micro-Python actually helps to decouple between the actual hardware but I have listed both the physical pin as well as the GPIO pin name.

Desc	Physical pin green	GPIO pin green	Physical pin red	GPIO pin red
Left eye	1	GP0	14	GP10
Right eye	2	GP1	15	GP11
Nose	4	GP2	16	GP12
Smile 1 (left up)	5	GP3	17	GP13
Smile 2 (right up)	6	GP4	19	GP14
Smile 3	7	GP5	20	GP15
Smile 4	9	GP6	21	GP16
Smile 5	10	GP7	22	GP17
Smile 6	11	GP8	24	GP18
Smile 7	12	GP9	25	GP19
Smile 8 (left down)	26	GP20	29	GP22
Smile 9 (right down)	27	GP21	31	GP26

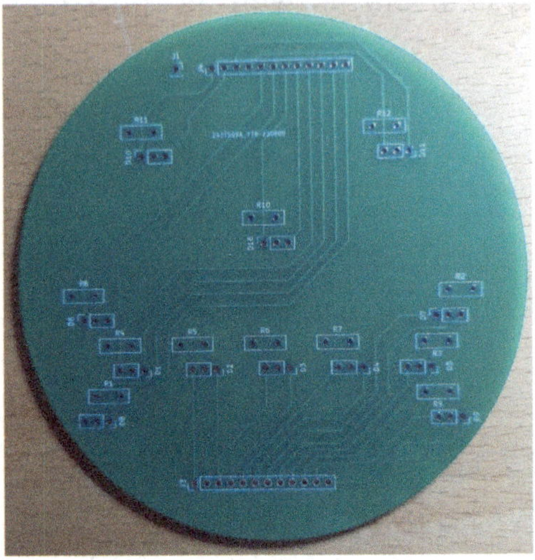

Figure 22: Smiley toy v2 board

Lessons learned

The second version of the smiley face should be easy enough to wire-up with all the connections centralized in these two spots. However, connecting up this many wires still does make for a rather bulky gadget. Round is a good shape for a face but it is rather difficult to get this to stand up and to display the face especially with all the wires connected. The Pico is typically powered via USB cable, but this smiley face doesn't have an on/off switch.

If you are creating your own printed circuit boards it is important to remember one rule.

Faster development
is not always better

This rule might seem a bit non-sequitur until you realize that the version 2 of the smiley toy is not very practical. It is super easy to design a circuit board and immediately upload it to a PCB board manufacturer. Yet unlike software there are a few notable differences when creating subsequent improved versions of PCB's.

Requires a new board
costs money
takes time to create
(more time to ship)
stuck with old boards

Perhaps a better rule to keep in mind might be:

Software is more malleable than hardware

Smiley toy V3
The second version of the smiley face looks good but it wasn't practical. Additional time was spent redesigning it so it can be set on a desk. The actual prototype, v1, pointed out how unmanageable wiring up a smiley face is. The v2 version of this toy also made it obvious that a completely round face is hard to use due to the number of wired connections. The ideal solution would have no wires. This can be implemented by attaching the Pico directly to the PCB.

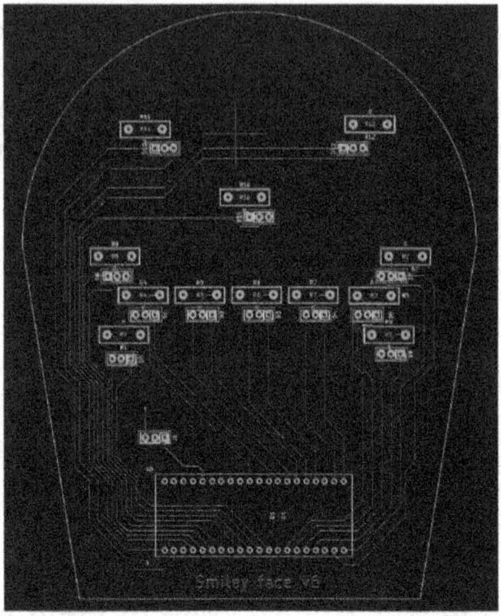

Using dual color leds

The one complication of using a dual color led is that it is really a combination of two leds in one package and has either two separate anodes and one common cathode or two separate cathodes and one common anode.

The ideal dual color led, in my opinion, would have two anodes and a common cathode. This would allow me to wire the common cathode directly to ground. Simply providing power to either of the anodes would then light up that color on the led.

The alternative dual color led is slightly different. Instead you provide power to the common anode and set one of the cathodes to ground. This sounds pretty obvious but the less obvious point is that the second cathode must simultaneously not be tied to ground. Both style dual color leds can be used in projects but the logic feels more natural when using a one with common cathode.

For version 3 of this project I am using dual color leds with a common anode. In order to test controlling a single led I wrote a small program, Listing 11, and wired up the Pico with a single led.

```
import time
from machine import Pin

LedAnode = Pin(11, Pin.OUT)
LedGreen = Pin(12, Pin.OUT)
LedRed   = Pin(13, Pin.OUT)

LedAnode.on()

# red on, green off
LedGreen.on()
LedRed.off()

# sleep 1 second
time.sleep_ms(1000)

# green on, red off
LedGreen.off()
LedRed.on()
```

Listing 11 Dual LED test code

As previously described, in order to turn on the red side of the dual color led we would need to set that red cathode to ground while simultaneously providing power to the anode. How can this be done while not setting the green cathode to ground?

This is done on the Pico by setting the green cathode to high and the red cathode to low. Current flows from high potential to low potential so the current flows through from the anode (pin 11) through the cathode of the red led (pin 13) and causes the led to light up red. Because the cathode of the green led (pin 12) is set to high no current will flow and thus the green led will not light up. This logic works fine but it is fairly counter intuitive when looking at the actual code.

```
def smileneutralred():
 smile1Red.off();  time.sleep_ms(smilepause)
 smile2Red.off();  time.sleep_ms(smilepause)
 smile3Red.off();  time.sleep_ms(smilepause)
 smile4Red.off();  time.sleep_ms(smilepause)
 smile5Red.off();  time.sleep_ms(smilepause)
```

This method “smileneutralred” actually turns on the red smile.

In addition to simply turning on the led, there is an additional pause that has been added. This causes the leds to be turned on one at a time in succession. If the pause is set small enough then the leds all turn on essentially instantaneously. Despite this looking like a complete success there was some difficulties.

Figure 23: Green frowning face

All of the faces using the red leds turned out just fine but there were some difficulties with the green leds on the smiley toy v3. It is somewhat difficult to see from Figure 19 but all of the proper leds are lit. The right eye and nose are barely light up. There can be a number of reasons for this.

Low quality led
Led requires more power
Problem with the resistor
Too large of a resistor

Imperfections of PCB
Bad soldering job

It is possible the leds that were acquired from a nameless organization on the internet were not high enough quality. It is just as likely that the green led requires more power than the Pico is able to provide. One of the problems of the discount leds that I am using is that they came without a specification sheet.

8 DEVELOPING WITH C/C++

If you have have been around an IT shop you may have already met someone who used punch cards or some other archaic editor for creating his or her programs. The editor may sound like an eye chart (emacs or vi) but the net result is that the developer did virtually all of the work without much support from the computer itself.

The good news is that in the intervening decades we have moved from a few command line tools and simple editors to an integrated development environment. This might mean syntax highlighting but generally includes compiling, debugging, but sometimes integration with source code control or other tools.

The number of IDE's available is pretty amazing. Some are limited to a specific language while others support many different languages and can be configured for your favorite language.

Geany - Integrated Development Env

The Geany IDE is preinstalled on the Raspberry Pi and is an easy way to get into C/C++ development.

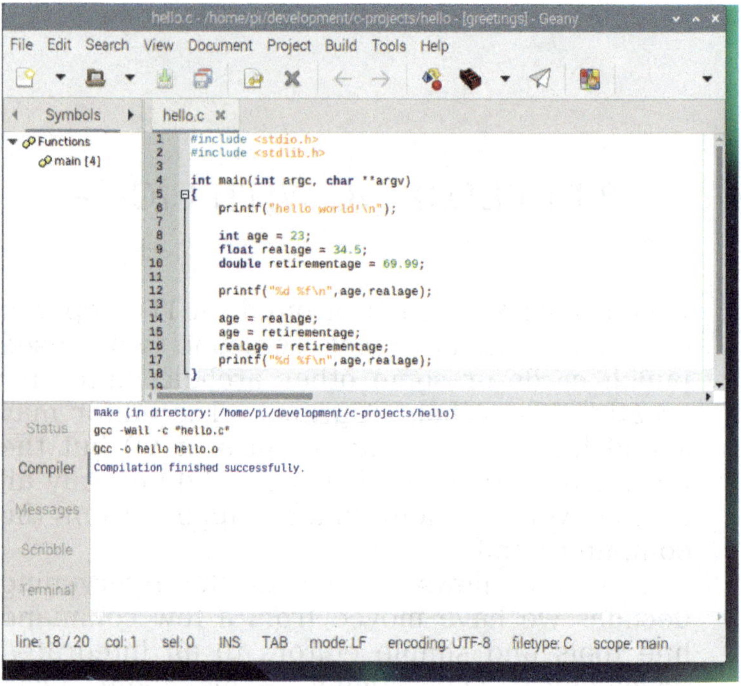

Geany has some tool icons for what will be undoubtedly the most common tasks.

Icon	Description
	New file
	Open file
	Save file
	Save all files
	Close file
	Compile the current source file
	Build will create the executable

Once you have entered some code you can click on compile or build to turn the code into object code or even an executable. There is a small difference between the two.

This example program consists of a single file. The compile method will compile the source code for the current into its object code. The build command is needed to compile and link the code into its final executable.

This is all pretty transparent for the user for

single source file programs. Pressing on the build button will perform the compile and link and you will have an executable. The IDE will create your executable without any "additional configuration".

However, for multiple source file programs a bit more configuration is needed. Traditionally C programs, on Unix or Linux, will have a Makefile.

A Makefile is the configuration file used by the make utility. This configuration defines the dependencies between all files in the project. A Makefile can contain many different rules, with these rules being for a single object or for the entire program. These rules can actually be used to run other programs to help with the software development process.

Makefile
```
all:        hello   calc

calc:
	gcc -Wall -o calc calc.c

hello:
	gcc -Wall -c hello.c
	gcc -Wall -c communicate.c
	gcc -o hello hello.o communicate.o

clean:
	rm hello.o communicate.o
	rm hello calc

test:       calc hello
	./calc
	./hello
```

This particular Makefile will create two different programs. The "calc" rule is perhaps the easiest to understand. The gcc compiler will create the calc executable from the calc.c source file. The compiler will perform a compile and link in a single step. The output of this rule is a program called "calc".

The hello rule is more complex. The hello executable could also be created in a single step similar to calc, but this example shows how to compile two different source files into object files with a final link to create the "hello" executable[24].

The typical hello world program is not complex and usually easy to debug even if you make a mistake. As programs get more complex, the mistakes become harder to spot you may need additional assistance and Geany IDE already has you covered.

Lint

Lint is a static code analysis tool used to flag programming errors, bugs, stylistic errors and suspicious constructs. This program lint was first created in 1978 by Stephen C Johnson at Bell laboratories.

This tool will read the source code of your C programs and will notify you of what it finds as syntactically suspicious. It might not be possible to write perfect code with no warnings or errors but ignore them at your peril. The warnings and errors only take a few moments to fix when they first appear. Not having any warnings will make newly introduced errors stand out all the more, and it is unlikely that

[24] The complete source for this example is in the appendix.

the development team will fix all of the messages at the end of the project. Taking time to remove these as they are encountered is time well spent.

Geany has a menu choice for running Lint. Any decent IDE will not hard code your tool chain nor does Geany. The build menu has a choice "set build commands" which allows you to setup which commands are used for each tool, Figure 24.

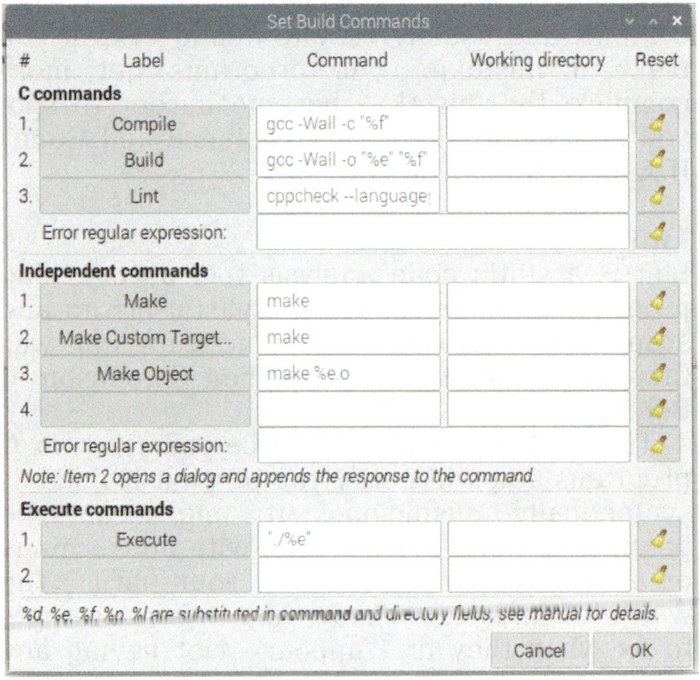

Figure 24: Configuring external tools in Geany

The lint that was already configured for

Geany is named cppcheck. This was not installed on my Raspberry Pi host but was installed from the package manager with the following command.

```
$ sudo apt-get install cppcheck
```

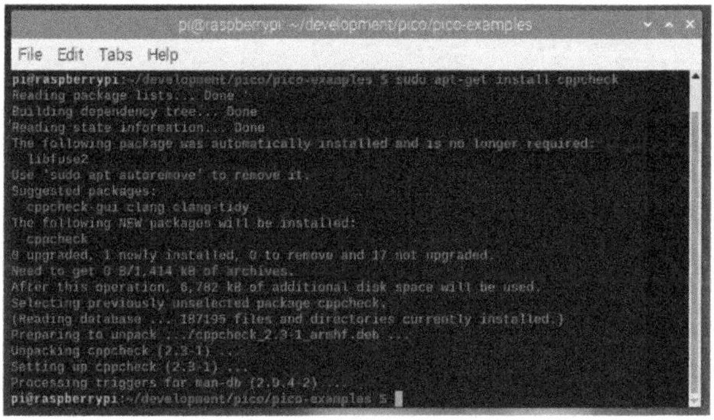

Figure 25: command to install cppcheck

The cppcheck tool does not have an icon on the toolbar but it can be selected from the build menu. When running this command, the analysis will be displayed in the window below the code, Figure 26.

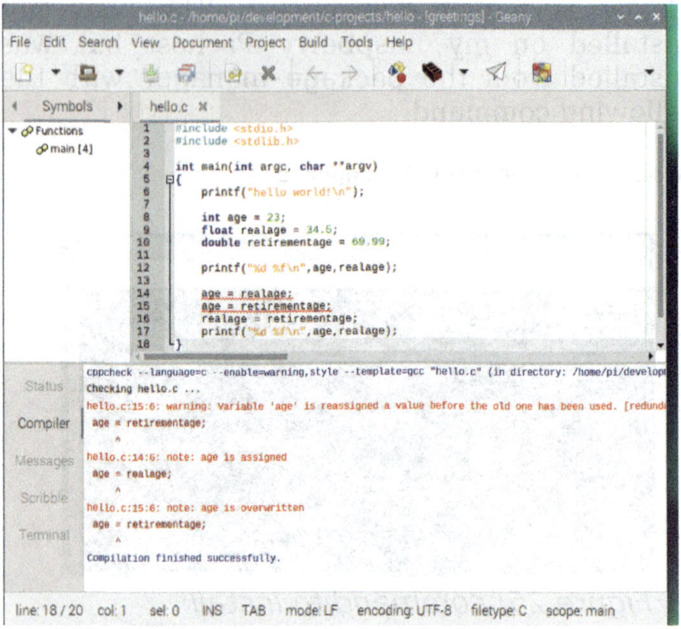

Figure 26: Geany running cppcheck tool

First C Program (Non Pico)

Despite its reputation, writing regular C programs is not terribly difficult, and in some ways perhaps even easier than writing for a Raspberry Pi Pico.

```
#include <stdio.h>
int main(int argc, char **argv)
{
  printf("Hello world!\n");
}
```

Listing 12, Standard hello world

The gcc compiler will turn this code into an executable for a personal computer.

```
$ gcc -Wall -o hello hello.c
```

Most real world programs are more complex than our simple hello world example and require either some C libraries or perhaps libraries from a framework. If you fail to provide required libraries your compiler / link step will complain with some error.

```
gcc -Wall -o calc calc.c
/usr/bin/ld: /tmp/cccOKsCJ.o: in function `area':
calc.c:(.text+0x2a): undefined reference to `floor'
collect2: error: ld returned 1 exit status
```

Listing 13 example linking error

This example, Listing 8, demonstrates that the linker could not find the floor function that was defined in the math library. Adding the math library to the compile step is quite simple. The -l command line argument with the library name will add this when linking. The math library is named libm.a and to link in the math library adding -lm will include libm.a.

```
gcc -Wall -o hello hello.c -lm
```

The process of linking in libraries is the same regardless of if you are compiling on your Windows PC or cross compiling for the Pico. The required SDK for the Pico will use its own libraries and will actually need a Makefile to build the executable along with the required libraries.

First Pico C Program

Most programming languages have a simple

"hello world" program that outputs some text, Listing 12. This is a bit tougher for platforms that don't have a screen.

The micro-controller version of this program does the closest possible thing which is to turn a LED on and off repeatably.

```
#include "pico/stdlib.h"

int main() {
    uint LED_PIN = 25;
    gpio_init(LED_PIN);
    gpio_set_dir(LED_PIN, GPIO_OUT);
    while (true) {
        gpio_put(LED_PIN, 1);
        sleep_ms(500);
        gpio_put(LED_PIN, 0);
        sleep_ms(500);
    }
```

Listing 14 Blink program written in C

This program calls four different functions that will not be found in your standard "getting started with C" programming book.

Function	Description
gpio_init	Initializes the pin so it can be used.
gpio_set_dir	This call sets the direction of the pin, to either input or output.
gpio_put	This pulls the pin to a logical 1 or 0.
sleep_ms	Sleep for a given number of milliseconds.

Looking at the code and the functions it calls makes it fairly obvious what the program does. One pin, number 25, will be essentially turned on and off every half second.

Figure 27: Pico onboard led and boot select button

In the appendix is a list of all of the pins and what features they can be used for such as UART or I2C but pin 25 isn't listed. It turns out that pin 25 is connected to an onboard led, Figure 26, surrounded by a yellow square. Near the onboard led is the "boot select" button, Figure 26, surrounded by an orange square. More information about the pinout of the Pico can be found in the appendix.

This small program requires the Pico SDK which is built to create the necessary library. Running the pico_setup.sh script in the previous chapter actually prepares your Raspberry Pi but it also downloads and builds the Pico SDK. This can be downloaded from github with the following command.

```
$ git clone https://github.com/raspberrypi/pico-sdk \
      /home/pi/development/pico-sdk
```

The pico_setup.sh actually provides some different examples with blink, Listing 14, being one of them. The blink example actually also contains one additional support file called CMakeLists.txt. This file contains some instructions for the build system. The build system is cmake which takes platform independent files, CMakeLists.txt, and will create native makefiles.

```
cmake_minimum_required(VERSION 3.12)

# comes from pico-sdk/external
include (pico_sdk_import.cmake)

project (blink C CXX ASM)
set(CMAKE_C_STANDARD 11)
set(CMAKE_CXX_STANDARD 17)

set(PICO_EXAMPLES_PATH ${PROJECT_SOURCE_DIR})

pico_sdk_init()
add_executable(blink blink.c)
target_link_libraries(blink pico_stdlib)
pico_add_extra_outputs(blink)
```

Listing 15 Makefile

This config file includes a file called pico_sdk_import.cmake which will contain more configuration. This file exists and is part of the pico_sdk project. Simply copy this file from the pico-sdk/external directory which was previously cloned using git.

In order to build our blink program you need the source file and cmake config file in a separate directory. You can either build the project in the same directory or in a separate (ie dist) directory.

1. mkdir -p /home/pi/development/blink
2. Create env variable PICO_SDK_PATH to Pico SDK directory
   PICO_SDK_PATH=/home/pi/development/pico-sdk
3. Change directory to blink
   cd /home/pi/development/blink
4. Create a build directory
   mkdir /home/pi/development/blink/dist
5. copy pico_sdk_import.cmake
   cp $PICO_SDK_PATH/external \
       /home/pi/development/blink
6. Change to dist directory
   cd /development/blink/dist
7. cmake
8. make

This set of instructions will create the blink.uf2 file that can be installed on the Pico.

Running first Pico program

The easiest way to install the program is to mount the Pico as a USB mass storage device.

Mounting a Pico is as simple as pressing the boot select button and holding it on the Pico when plugging in the USB cable to your Raspberry Pi until you are propted to open up new device.

Installing the program is then as simple a copying over the blink.uf2 file to this USB device. The program will be installed and the USB device will be unmounted. If the program is the blink program, Listing 14, the LED on the board next to the boot select will start flashing.

The Pico has now been programmed and can be disconnected from the Raspberry Pi. Each time from this point on when the Pico is powered up, this program will run.

Troubleshooting

This method of "flashing" a micro-controller with a program is really easy and almost foolproof. However, be careful when reaching for that USB cable. If there is either a problem with the cable or if it is power only, then the Pico won't mount as expected.

You can check out what diagnostic messages were generated during this process using the dmesg command in a terminal window.

```
$ dmesg
[  4370.000004] usb 1-1.3: new full-speed USB device number 11 using xhci hcd
[  4370.000283] usb 1-1.3: Device not responding to setup address.
[  4370.220253] usb 1-1.3: Device not responding to setup address.
[  4370.440024] usb 1-1.3: device not accepting address 11, error -71
[  4370.540048] usb 1-1.3: new full-speed USB device number 12 using xhci_hcd
```

[4370.540314] usb 1-1.3: Device not responding to setup address.
[4370.760279] usb 1-1.3: Device not responding to setup address.
[4370.980011] usb 1-1.3: device not accepting address 12, error -71
[4370.980681] usb 1-1-port3: unable to enumerate USB device

APPENDIX - SMILEY CODE V3

The complete listing of the Smiley code is below. This code is a bit repetitive. It could have been made shorter and more concise by having a data section with each block of data describing which leds should be lit. This could also have encoded which color should be used. However, that code would have been much harder to read. The goal isn't to teach data structures and algorithms but rather to show how to control a Raspberry Pi Pico.

```
import time
from machine import Pin

lefteyeRed = Pin(1, Pin.OUT)
lefteyeGreen = Pin(0, Pin.OUT)

righteyeRed = Pin(2, Pin.OUT)
righteyeGreen = Pin(3, Pin.OUT)

noseGreen = Pin(5, Pin.OUT)
```

```
noseRed = Pin(4, Pin.OUT)

smile1Red = Pin(26, Pin.OUT)
smile2Red = Pin(21, Pin.OUT)
smile3Red = Pin(19, Pin.OUT)
smile4Red = Pin(17, Pin.OUT)
smile5Red = Pin(10, Pin.OUT)
smile6Red = Pin(6, Pin.OUT) # up
smile7Red = Pin(15, Pin.OUT) # up
smile8Red = Pin(8, Pin.OUT) # down
smile9Red = Pin(12, Pin.OUT) # down

smile1Green = Pin(22, Pin.OUT)
smile2Green = Pin(20, Pin.OUT)
smile3Green = Pin(18, Pin.OUT)
smile4Green = Pin(16, Pin.OUT)
smile5Green = Pin(11, Pin.OUT)
smile6Green = Pin(7, Pin.OUT) # up
smile7Green = Pin(14, Pin.OUT) #up
smile8Green = Pin(9, Pin.OUT) # down
smile9Green = Pin(13, Pin.OUT) #down

# ms delay between turning each led on
smilepause = 100

# all off

def sleepTenSeconds():
 time.sleep_ms(10000)

def alloff():
 lefteyeRed.on()
 lefteyeGreen.on()
 righteyeRed.on()
 righteyeGreen.on()
 noseRed.on()
 noseGreen.on()

 smile1Red.on()
 smile2Red.on()
 smile3Red.on()
 smile4Red.on()
 smile5Red.on()
```

```
 smile6Red.on()
 smile7Red.on()
 smile8Red.on()
 smile9Red.on()

 smile1Green.on()
 smile2Green.on()
 smile3Green.on()
 smile4Green.on()
 smile5Green.on()
 smile6Green.on()
 smile7Green.on()
 smile8Green.on()
 smile9Green.on()

# red faces support routines

def eyesred():
 lefteyeRed.off()
 righteyeRed.off();

def nosered():
 noseRed.off()

def smilered():
 smile6Red.off();
 time.sleep_ms(smilepause)
 smileneutralred()
 smile7Red.off();
 time.sleep_ms(smilepause)

def smileneutralred():
 smile1Red.off();
 time.sleep_ms(smilepause)
 smile2Red.off();
 time.sleep_ms(smilepause)
 smile3Red.off();
 time.sleep_ms(smilepause)
 smile4Red.off();
 time.sleep_ms(smilepause)
 smile5Red.off();
 time.sleep_ms(smilepause)
```

```
def frownred():
 smile8Red.off();
 time.sleep_ms(smilepause)
 smileneutralred()
 smile9Red.off();
 time.sleep_ms(smilepause)

# green faces support routines

def eyesgreen():
 lefteyeGreen.off()
 righteyeGreen.off();

def nosegreen():
 noseGreen.off()

def smilegreen():
 smile6Green.off();
 time.sleep_ms(smilepause)
 smileneutralgreen()
 smile7Green.off();
 time.sleep_ms(smilepause)

def smileneutralgreen():
 smile1Green.off();
 time.sleep_ms(smilepause)
 smile2Green.off();
 time.sleep_ms(smilepause)
 smile3Green.off();
 time.sleep_ms(smilepause)
 smile4Green.off();
 time.sleep_ms(smilepause)
 smile5Green.off();
 time.sleep_ms(smilepause)

def frowngreen():
 smile8Green.off();
 time.sleep_ms(smilepause)
 smileneutralgreen()
 smile9Green.off();
 time.sleep_ms(smilepause)
```

```
# methods for full faces

def redfrownyface():
 alloff()
 eyesred()
 time.sleep_ms(smilepause)
 nosered()
 time.sleep_ms(smilepause)
 frownred()

def greenfrownyface():
 alloff()
 eyesgreen()
 time.sleep_ms(smilepause)
 nosegreen()
 time.sleep_ms(smilepause)
 frowngreen()

def redhappyface():
 alloff()
 eyesred()
 time.sleep_ms(smilepause)
 nosered()
 time.sleep_ms(smilepause)
 smilered()

def greenhappyface():
 alloff()
 eyesgreen()
 time.sleep_ms(smilepause)
 nosegreen()
 time.sleep_ms(smilepause)
 smilegreen()

def greenneutralface():
 alloff()
 eyesgreen()
 time.sleep_ms(smilepause)
 nosegreen()
 time.sleep_ms(smilepause)
 smileneutralgreen()

def redneutralface():
```

```
 alloff()
 eyesred()
 time.sleep_ms(smilepause)
 nosered()
 time.sleep_ms(smilepause)
 smileneutralred()

# initialize the current state
alloff()

#start to display different expressions
for i in range(5):

 greenhappyface()
 sleepTenSeconds()

 greenneutralface()
 sleepTenSeconds()

 greenfrownyface()
 sleepTenSeconds()

 redfrownyface()
 sleepTenSeconds()

 redneutralface()
 sleepTenSeconds()

 redhappyface()
 sleepTenSeconds()

# all off once we are done.
alloff()
```

Listing 16 Smiley v3 driver code

10 APPENDIX - KEYPAD EXAMPLE

Here is the full listing of the "complex button" example described in chapter 5.

```
import time
from machine import Pin

# global variables - start

# arrays of pins connected directly to keypad
column_pins = []
row_pins = []

# entered keys
keybuffer = []

# secret passcode
passcode = "12345"

redled = Pin(15, Pin.OUT, Pin.PULL_UP)
greenled = Pin(14, Pin.OUT, Pin.PULL_UP)

# global variables - end

#
# things that will not change (ie wiring setup)
#
keyarray = [
    ['1', '2', '3', 'A'],
    ['4', '5', '6', 'B'],
```

```
    ['7', '8', '9', 'C'],
    ['*', '0', '#', 'D'],
    ]

# these are the pins of our pico
# connected to the keypad
keypad_rows = [ 16,17,18,19]
keypad_cols = [ 20,21,26,27]

# updating arrays of row and col pins
for idx in range(0,4): # 0 - 3
    row_pins.append(Pin(keypad_rows[idx], Pin.OUT))
    column_pins.append(Pin(keypad_cols[idx], Pin.IN,
Pin.PULL_DOWN))


# initialize their values
for idx in range(0,4): # 0 - 3
    row_pins[idx].value(1)
    column_pins[idx].value(0)

def getkey():
    retkey = None
    done = False

    while done == False:

        for rowidx in range(0,4):
            for colidx in range(0,4):
                row_pins[rowidx].high()
                if column_pins[colidx].value() == 1:
                    retkey = keyarray[rowidx][colidx]
                    print("found one, ",retkey)
                    done = True
                    time.sleep_ms(500)
                row_pins[rowidx].low()


        time.sleep_ms(100)

    return retkey

def getline():
    done = False
    keybuffer = []
    val = None
    while done == False:
        val = getkey()
        if val == '#':
            done = True
```

```
        else:
            keybuffer.append(val)

    # return as a string
    retval = ""
    for single in keybuffer:
        retval += single

    print("retval line = " + retval)
    return retval


def main():

    # you shalt not pass
    redled.on()
    greenled.off()

    # get the secret passcode
    done = False
    while done == False:
        val = getline()
        if val == passcode:
            print("success")
            done = True
        else:
            print(val)

    # eh, ok, go on in.
    greenled.on()
    redled.off()
    time.sleep_ms(2000)

main()

greenled.off()
redled.off()
```

11 APPENDIX - MAKEFILE / C

This source is not really important for the pico, but it is included for completeness.

calc.c

```
#include <stdio.h>
#include <stdlib.h>

int main(int argc, char **argv) {

   int a = 1, b = 2;
   int total = a + b;
   printf("total of %d + %d = %d \n",a,b,total);
}
```

Listing 17 Single source c program

communicate.c

```
#include <stdlib.h>
#include <stdio.h>
void communicate(char *text) {

      printf("%s\n",text);
}
```

Listing 18 source file with a function

communicate.h

```
void communicate(char *text) ;
```

Listing 19 Header file for communicate function

hello.c

```
#include <stdio.h>
#include <stdlib.h>
#include "communicate.h"

int main(int argc, char **argv) {

	char msg[] = "hello world!\0";
	communicate(msg);
}
```

Listing 20 main file - multi source file program

Makefile

```
all:	hello	calc

calc:
	gcc -Wall -o calc calc.c

hello:
	gcc -Wall -c hello.c
	gcc -Wall -c communicate.c
	gcc -o hello hello.o communicate.o

clean:
	rm hello.o communicate.o
	rm hello calc

test:	calc hello
	./calc
	./hello
```

Listing 21 Makefile

12 APPENDIX - PIN FUNCTIONS

Left side of Pico pins, reference chart

UART0 TX	I2C0 SDA	SPI0 RX	GP0	1
UART0 RX	I2C0 SCL	SPI0 Csn	GP1	2
			GND	3
	I2C1 SDA	SPI0 SCK	GP2	4
	I2C1 SCL	SPI0 TX	GP3	5
UART1 TX	I2C0 SDA	SPI0 RX	GP4	6
UART1 RX	I2C0 SCL	SPI0 Csn	GP5	7
			GND	8
	I2C1 SDA	SPI0 SCK	GP6	9
	I2C1 SCL	SPI0 TX	GP7	10
UART1 TX	I2C0 SDA	SPI1 RX	GP8	11
UART1 RX	I2C0 SCL	SPI1 Csn	GP9	12
			GND	13
	I2C1 SDA	SPI1 SCK	GP10	14
	I2C1 SCL	SPI1 TX	GP11	15
UART0 TX	I2C0 SDA	SPI1 RX	GP12	16
UART0 RX	I2C0 SCL	SPI1 Csn	GP13	17
			GND	18
	I2C1 SDA	SPI1 SCK	GP14	19
	I2C1 SCL	SPI1 TX	GP15	20

Power
Ground
UART
UART default
GPIO, PIO, and PWM
ADC
SPI
SPI default
I2C
I2C default
System control

Right side of Pico pins, reference chart

40	VBUS			
39	VSYS			
38	GND			
37	3V3_EN			
36	3V3_OUT			
35		ADC_VREF		
34	GP28	ADC2		
33	GND	AGND		
32	GP27	ADC1	I2C1 SCL	
31	GP26	ADC0	I2C1 SDA	
30	RUN			
29	GP22			
28	GND			
27	GP21		I2C0 SCL	
26	GP20		I2C0 SDA	
25	GP19	SPI0 TX	I2C1 SCL	
24	GP18	SPI0 SCK	I2C1 SDA	
23	GND			
22	GP17	SPI0 Csn	I2C0 SCL	UART0 TX
21	GP16	SPI0 RX	I2C0 SDA	UART0 RX

- Power
- Ground
- UART
- UART default
- GPIO, PIO, and PWM
- ADC
- SPI
- SPI default
- I2C
- I2C default
- System control

ABOUT THE AUTHOR

Christopher was born in the United States in the Midwest and started his career as a software developer but apparently became restless and made the move to Europe. He has spent most of his time working in IT departments for banks and large corporations, usually as a consultant but occasionally as a software developer. His work brought him in contact with different people from different cultures but typically in a workplace environment.

It was during one of the IT projects he was lucky enough to meet an old time electrical engineer from the former Soviet Union. During the daytime we worked on software and in the evenings he infused me with the joy of hardware creation.

Made in the USA
Monee, IL
03 November 2026

67380149R00075